Application Lifecycle Management on Microsoft Power Platform

A comprehensive guide to managing the deployment of your solutions

Benedikt Bergmann

‹packt›

Application Lifecycle Management on Microsoft Power Platform

Associate Group Product Manager: Aaron Tanna
Publishing Product Manager: Kushal Dave
Book Project Manager: Manisha Singh
Senior Content Development Editor: Rosal Colaco
Technical Editor: Vidhisha Patidar
Copy Editor: Safis Editing
Indexer: Tejal Soni
Production Designer: Ponraj Dhandapani
Business Development Executive: Kritika Pareek

First published: December 2024

Production reference: 1060924

Published by Packt Publishing Ltd.
Grosvenor House
11 St Paul's Square
Birmingham
B3 1RB, UK

ISBN 978-1-83546-232-4

www.packtpub.com

Foreword

When I first met Benedikt Bergmann, before he was awarded Microsoft Most Valuable Professional, I remember being impressed by the presentation he was delivering on advanced Application Lifecycle Management (ALM) within Power Platform. Since then, I've had the pleasure of calling him a friend and peer, witnessing him become one of the leading experts in the field of Application Lifecycle Management and Power Platform solution development. Just as Benedikt's presentations are clear, insightful, and well thought out, in the same way, his latest book will guide you through the basics as well as some of the more complex aspects of ALM and its application within Power Platform.

Do you have a source code control and environment strategy? Do you know how to effectively segment your solutions? This book explores these topics and more, bringing you techniques and opinions that Benedikt has collected through his years of experience as a Power Platform consultant. The content is delivered in an accessible and engaging way, using examples, well-laid-out step-by-step instructions, screenshots, YAML code snippets, and links to additional sources of information. You will learn how to structure your solutions, create quality gates, develop a branching strategy, and create automated build and release pipelines.

I believe that Benedikt's book is a valuable resource for anyone looking to understand and implement ALM, perhaps one of the most overlooked aspects of Power Platform solution delivery. By the end of this book, you'll not only grasp the theoretical aspects of ALM within Power Platform but also be ready to apply them to real-world projects. Start reading today, and let Benedikt help you raise the quality bar of your next Power Platform project.

- Scott Durow, Microsoft Cloud Developer Advocate

Contributors

About the author

Benedikt Bergmann is the CEO of CRM Konsulterna Deutschland and develops solutions for Dynamics 365 and the Power Platform with his German precision and nearly 15 years of experience. He is a solution architect with a huge understanding of customers' challenges and a real problem solver with very detailed knowledge about most of the Platform.

One area he is very invested in is everything around Application Lifecycle Management. In 2021, he was awarded Microsoft Most Valuable Professional in the Business Application area.

I would like to thank my loving and patient wife, for her continued support, patience, and encouragement throughout the long process of writing this book.

About the reviewer

Wael Hamze is a Power Platform Lead and a 7 times Microsoft Business Applications MVP with over 20 years of experience in software design and development. He is an expert in Microsoft technologies with a focus on Power Platform and Dynamics 365. Wael is a passionate speaker, author, and mentor in the Power Platform community. Wael is the author of "Power DevOps Tools", which is the first widely adopted Azure DevOps extension for Power Platform that has been adopted by thousands of organizations to implement automated DevOps processes for Power Platform and Dynamics 365. His mission is to enable and empower teams and organizations to achieve their goals and optimize their performance by providing simple innovative solutions.

Table of Contents

5

Power Platform CLI   55

6

Environment Variables, Connection References, and Data   67

7

Approaches to Managing Changes in Power Platform ALM   79

11

Advanced Techniques   149

12

Continuous Integration/Continuous Delivery   167

13

Deepening ALM Knowledge   183

Assessments   189

Index   191

Other Books You May Enjoy   200

Preface

Application Lifecycle Management (**ALM**) is a crucial aspect of software development. It encompasses the entire lifecycle of an application, from conception to decommissioning, and includes automated deployment to downstream environments. The different stages of a complete ALM process are:

- Plan
- Develop
- Test
- Deploy
- Maintain

This book will primarily focus on the automated deployment aspect.

In recent months, there has been a surge in attention towards ALM in the Power Platform. This heightened interest is driven by Microsoft's efforts to enhance the Platform and by the desire of customers and implementers to achieve more reliable and predictable deployments.

To succeed in this endeavor, it is essential to have a strong grasp of Power Platform fundamentals, such as Environments, Solutions, and Power Platform CLI, as well as a deep understanding of the various ALM options available. This book will comprehensively cover these areas and provide practical examples of automated deployment implementations.

Who this book is for

The first part of the book is intended for anyone working with the Power Platform or Dynamics 365. It is crucial to grasp the concept of Application Lifecycle Management and its implications.

The second half is geared towards technical individuals capable of implementing the described pipelines.

What this book covers

Chapter 1, An Intro to ALM, provides a general, unrelated to the Power Platform, overview of what the Application Lifecycle is.

Chapter 2, ALM in Power Platform, relates the general ALM approach to the Power Platform. This chapter also covers the usual problems one experiences when trying to implement a proper ALM process for the Power Platform.

Chapter 3, Power Platform Environments, will explain in detail what is meant when talking about Environments when it comes to the Power Platform. We will discuss different environment setups.

Chapter 4, Dataverse Solutions, deepens your knowledge about solutions in Power Platform. We learn what they are for, and not for, as well as different approaches to structure them.

Chapter 5, Power Platform CLI, explains everything one needs to know about the Power Platform CLI when it comes to ALM. We will mostly focus on the commands around Solutions.

Chapter 6, Environment Variables, Connection References, and Data, dives into areas to make implementation more dynamic and how to deploy those components securely and predictably to downstream environments.

Chapter 7, Approaches to Managing Changes in Power Platform ALM, explains the difference between environment-centric and source code-centric approach. We also will learn about branching, and which pipelines a source code-centric approach requires.

Chapter 8, Essential ALM Tooling for Power Platform, dives into the different tools we have at hand to implement an ALM process for the Power Platform and how to select the correct one for your project.

Chapter 9, Project Setup, goes through everything that needs to be set up to implement the previously discussed approaches. This includes creating Service Principal Names (SPN), setting up Azure DevOps, setting up GitHub, and setting up Power Platform Pipelines. It also briefly explains how project folders could be structured.

Chapter 10, Pipelines, is the chapter where the magic happens, and we learn how to implement the source code-centric approach.

Chapter 11, Advanced Techniques, the very basic implementation we do in *Chapter 10* usually requires some extension to fit a real-life scenario. This chapter will explain some of them. We will learn about advanced YAML (variables, parameters, conditions, and loops), settings files, healthy code state, and transporting data.

Chapter 12, Continuous Integration/Continuous Delivery, looks at which areas you need to focus on to implement a proper CI/CD process for the Power Platform. We will learn more about branching, quality gates, Pull requests, and Versioning.

Chapter 13, Deepening ALM Knowledge, is the last chapter of this book and focuses on which areas you could dive more into after finishing this book.

To get the most out of this book

- Azure DevOps or GitHub Account

- Two Dataverse environments (one for source and one for target)

- App Registration: follow this instruction: `https://benediktbergmann.eu/2022/01/04/setup-a-service-principal-in-power-automate/`

If you are using the digital version of this book, we advise you to type the code yourself or access the code from the book's GitHub repository (a link is available in the next section). Doing so will help you avoid any potential errors related to the copying and pasting of code.

Download the example code files

You can download the example code files for this book from GitHub at `https://github.com/PacktPublishing/Application-Lifecycle-Management-on-Microsoft-Power-Platform`. If there's an update to the code, it will be updated in the GitHub repository.

We also have other code bundles from our rich catalog of books and videos available at `https://github.com/PacktPublishing/`. Check them out!

Conventions used

There are a number of text conventions used throughout this book.

`Code in text`: Indicates code words in text, database table names, folder names, filenames, file extensions, pathnames, dummy URLs, user input, and Twitter handles. Here is an example: "In the `PowerPlatformSPN` parameter, you need to provide the name of the service connection to the development environment."

A block of code is set as follows:

```
- task: PowerPlatformPublishCustomizations@2
  displayName: Publish Customizations
  inputs:
    authenticationType: 'PowerPlatformSPN'
    PowerPlatformSPN: '<Name of the dev ADO Service connection>'
    AsyncOperation: true
    MaxAsyncWaitTime: '60'
```

Bold: Indicates a new term, an important word, or words that you see onscreen. For instance, words in menus or dialog boxes appear in **bold**. Here is an example: "Within the GitHub project in question, we navigate to **Actions** and select the **New workflow** button."

> **Tips or important notes**
> Appear like this.

Get in touch

Feedback from our readers is always welcome.

General feedback: If you have questions about any aspect of this book, email us at customercare@packtpub.com and mention the book title in the subject of your message.

Errata: Although we have taken every care to ensure the accuracy of our content, mistakes do happen. If you have found a mistake in this book, we would be grateful if you would report this to us. Please visit www.packtpub.com/support/errata and fill in the form.

Piracy: If you come across any illegal copies of our works in any form on the internet, we would be grateful if you would provide us with the location address or website name. Please contact us at copyright@packt.com with a link to the material.

If you are interested in becoming an author: If there is a topic that you have expertise in and you are interested in either writing or contributing to a book, please visit authors.packtpub.com.

Share Your Thoughts

Once you've read *Application Lifecycle Management on Microsoft Power Platform*, we'd love to hear your thoughts! Scan the QR code below to go straight to the Amazon review page for this book and share your feedback.

https://packt.link/r/1835462324

Your review is important to us and the tech community and will help us make sure we're delivering excellent quality content.

Download a free PDF copy of this book

Thanks for purchasing this book!

Do you like to read on the go but are unable to carry your print books everywhere?

Is your eBook purchase not compatible with the device of your choice?

Don't worry, now with every Packt book you get a DRM-free PDF version of that book at no cost.

Read anywhere, any place, on any device. Search, copy, and paste code from your favorite technical books directly into your application.

The perks don't stop there, you can get exclusive access to discounts, newsletters, and great free content in your inbox daily

Follow these simple steps to get the benefits:

1. Scan the QR code or visit the link below

https://packt.link/free-ebook/9781835462324

2. Submit your proof of purchase
3. That's it! We'll send your free PDF and other benefits to your email directly

1

An Intro to ALM

The most productive companies deploy custom software on a regular schedule. To achieve this kind of efficiency, a flawless process of software development from start to end is crucial. This is where **application lifecycle management (ALM)** comes into the picture.

ALM adoption, especially within the Power Platform, is growing. There are a lot of questions people have about this topic.

This first chapter will give you a general introduction to ALM, as well as what is included.

In this chapter, we're going to cover the following main topics:

- What is ALM?
- What are the stages of an ALM process?
- What the benefits of ALM are and why we should use it
- The tools of an ALM strategy

By the end of this chapter, you will be familiar with the key concepts and ideas in ALM. The topics covered will help you understand what ALM contains and why it is important.

ALM overview

When we talk about ALM, we often only talk about the automation part. Even though this is an essential part of it, there is more to ALM than that. ALM is a complex system of tools, people, and processes to control the complete cycle of an application, from planning and development, testing, and maintenance to retirement. It is important to know that all of these are integral parts of holistic ALM.

The following diagram illustrates the different stages of an ALM process and that it is an ever-repeating cycle:

Figure 1.1 – Stages of ALM

Whenever we have completed one cycle by starting from the **Maintain** stage, we start over with the planning of the next iteration. After the first iteration, the **Maintain** stage continues while the other stages are executed to support and maintain the current version in production while the next version is prepared. This goes on until the application is finally retired.

Stages of the ALM cycle

As you can see in *Figure 1.1*, the ALM cycle contains five stages:

1. **Plan**
2. **Develop**
3. **Test**
4. **Deploy**
5. **Maintain**

The following sections will explain these five stages.

Plan

First, an application must be planned. Therefore, this is the first stage of the ALM process. This will include requirement gathering from all stakeholders to collect all their needs for the application to support their business cases in the best way possible. Business requirements are not the only requirements we have to take into consideration. Within this stage, gathering compliance and governance requirements is as important as gathering business requirements. Usually, we differentiate between functional and non-functional requirements, where business requirements are often functional and the others mentioned are often non-functional.

With all this input and requirements, a design of the application will be created.

Since the ALM process is a cycle and can repeat itself, it can support an Agile development approach. In this case, the requirements gathered in the **Plan** stage of one iteration can change the requirements and design of an earlier iteration.

Develop

After the plan for the application (or the next iteration) is completed, the team starts implementing it. The development team must break down given requirements into small pieces to be able to implement them.

Within this stage, the team could use different development approaches. The most known ones are either a Waterfall or an Agile approach. Since the whole ALM process is a circle and repeats itself, this would mean that the chosen development approach is a sub-approach.

Test

Usually, the **Test** stage and the **Develop** stage are very tightly related to each other and even overlap. Testers start their work of defining test cases and setting up test environments while the application is still in development. They send features back to development whenever they find parts that are not working as expected, which should happen in close communication between the test team and the development team.

In addition to the often-manual tests of dedicated test teams, it is recommended that the development team also creates unit tests as well as integration tests for the software they are writing. In addition to that, there might be the need for automated UI tests, performance tests, load tests, and a bunch of other tests.

The formal purpose of this stage is to verify that the created application meets the requirements defined in the planning stage of the ALM process. Until this is the case, the team will do mini-iterations between the **Develop** and **Test** stages.

Deploy

This stage defines the release of the product to production or the end user. There is no direct definition of how this stage should be done. It very much depends on the kind of software developed, the target audience, the prerequisites within the organization, and a lot of other factors.

Maintain

The last stage is to maintain your application, as the name says. It includes monitoring and managing the application. This stage is usually the longest in an ALM process.

As soon as an application is in this stage, it stays there until it gets retired, even though new iterations (a.k.a., new versions) are usually started as soon as one version goes to production and therefore enters the **Maintain** stage.

Advantages of ALM

As mentioned earlier, a flowless process to develop custom software is crucial to the success of an implementation. In the upcoming sections, we will analyze the following key advantages of ALM:

- Boosting communication and strengthening visibility
- Sharpening testing
- Increasing quality
- Improving development and deployment speed

Boosting communication and strengthening visibility

A good ALM strategy provides your team with the tools needed to deliver high-quality software. This includes proper team communication and planning platform(s). All of the tools in an ALM strategy should be as integrated as they can be to improve collaboration.

When those tools and platforms are in place, all stakeholders, and the development team in particular, will get better insights into the current project status and what has been achieved already, as well as what lies ahead.

Sharpening testing

Another part of a sophisticated ALM strategy should be guidelines, best practices, and tools when it comes to testing. This includes manual tests such as regression tests by the test team, as well as automated tests such as, for example, unit tests or integration tests.

When those guidelines, best practices, and tools are in place, the development team can easily create automated tests, which should be run on a regular basis, at least before the software is deployed to the downstream environments.

Increasing quality

The first two key advantages together will lead to increased software quality. Since all stakeholders are able to communicate better and more frequently and get better insights into the current project status, and the development team can create automated tests, the overall quality of your delivered software will be increased, mainly because the identification of issues can be done a lot earlier in the process.

Through source code management and proper branching, the quality of the delivery can be increased even more.

Improving development and deployment speed

All of the aforementioned will, in the end, improve development and deployment speed. Since issues can be identified earlier in the process, the development team needs less time to fix them and respond to the input of the testers. Therefore, the cognitive shift between tasks becomes less for both developers and testers, which leads to more efficiency and thus improved productivity.

The deployment process that should be set up during the establishment of an ALM strategy should optimize and automate the current deployment process. Through iterations of improvements, it will be optimized even further. This means that the deployment speed will be increased over time.

ALM tools

ALM tools are usually a collection of project management tools. How sophisticated those tools are depends on a lot of factors, one of which is how mature the current organization is in relation to application development. It could range from a very simple document or wiki to a "full-blown" ALM product.

It is also important to mention that there are usually several tools that are combined to achieve a holistic ALM tool landscape.

The following list gives you an idea of the features your ALM tool landscape should (at least) cover:

- **Communication**: The team implementing the solution needs to communicate with various stakeholders. Since communication is a crucial part of developing an application, it should be part of the ALM planning. It should be clear to the entire implementation team which communication channels they should use.

- **Requirement planning**: Requirements are necessary so that the development team implements the correct functionalities according to the needs of the business. Therefore, clear structure and tooling in regard to requirements (including gathering, defining, estimation, and tracking) are important.

- **Test management**: Every implementation should be tested. An automated approach is preferable, but a manual approach in most cases is also feasible. Both approaches are often combined, depending on which area of the application is currently being tested. Either way, a good ALM approach should define where the tests are managed. This includes defining new tests, scheduling test execution, and tracking test results.

- **Source code management**: Since a usual application implementation project includes writing code, an ALM process has to include a tool to manage the project's source code. This includes different branches, commit tracking, and pull requests.

- **Automated deployment**: An optimized ALM process will execute deployments to downstream environments in an automated fashion. This optimizes quality since all the necessary steps are always executed in the same order and in the same way. To be able to handle this, an ALM process should define which deployment automation tool to use.

Here are some examples of ALM tools:

- Microsoft Azure DevOps
- GitHub Actions
- Atlassian Software Suite (Jira, Confluence, Trello, Bitbucket, Bamboo)
- Jama Software
- Tuleap
- Visure

Summary

In this chapter, we learned about the key ideas behind ALM. We learned that an ALM process contains five steps: plan, develop, test, deploy, and maintain.

This chapter also describes why ALM is important to a successful implementation project and the advantages it brings. They are as follows:

- Boosting communication and strengthening visibility
- Sharpening testing
- Increasing quality
- Improving development and deployment speed

You now understand the stages of ALM and some of the tools used to implement it.

The next chapter looks more deeply into the need for ALM when working with the Microsoft Power Platform.

Questions

1. Which of the following is a benefit of ALM?

    A. Code works the first time without errors

    B. Qualify of code is improved

    C. Code is easier to maintain

    D. Telemetry is automatically captured

2. Which two of the following ALM tools are provided by Microsoft?

   A. Azure DevOps

   B. GitHub

   C. Jira

   D. Tuleap

Further reading

For more details on the topics covered in this chapter, reference the following resources:

- *Application lifecycle management*: https://learn.microsoft.com/dynamics365/guidance/implementation-guide/application-lifecycle-management

- *DevOps solutions on Azure*: https://azure.microsoft.com/solutions/devops/

2

ALM in Power Platform

In the previous chapter, we learned about what **Application Lifecycle Management** (ALM) is in general and why it is important. This second chapter will go into detail about the following topics:

- Terminology specific to Power Platform
- Difference between "normal" ALM and Power Platform ALM
- Solution levels

By the end of this chapter, you will have a good overview of why ALM is important in a Power Platform project. You will also have learned some of the relevant terminology.

Terminology

First of all, we have to establish an understanding of some important terminology when it comes to Power Platform. Those terms will be used throughout the rest of the book.

Tenant

A **tenant** defines the boundaries of everything an organization is doing with Microsoft. It can contain Azure components, Power Platform components, SharePoint, and a lot more.

> **Multiple tenants**
>
> Some organizations do have multiple tenants. This can have several reasons, such as separation for dev, test, and prod, or because parts of the company were acquired, for example.

Environment

In Power Platform, an **environment** is a space where one can store, manage, and share data, as well as business logic in the form of, for example, apps, flows, or chatbots.

There can be several environments within a Microsoft tenant.

Organizations can choose to have different environments for different countries, Apps, departments, or stages in a development process. We will learn more about environments in *Chapter 3*.

Solution

A **solution** is a container to store all the customizations and configurations within an environment. Data cannot be added to solutions. Solutions are used to transport customizations to a downstream environment.

Every environment can have several solutions.

Component

Components are defined as everything that can be added to a solution. For example, tables, flows, connection references, canvas apps, or model-driven apps can be added.

Configuration

Configurations are all the changes one can make to an environment within the boundaries of functionality that we have out of the box, or without changing any existing components or creating new components. System settings, Server-side sync, or duplication detection rules are some examples of configuration.

Customization

Customization refers to changes to an environment that extend existing functionality or create new functionality. Flows, tables, canvas apps, and model-driven apps are some examples of customizations.

Development

With **development**, we usually mean everything that needs to be developed by a pro coder since it requires someone to write code. This can be, for example, plugins, custom actions, custom components (**Power Apps Component** Framework (**PCF**)), or web resources.

Solution-aware

As mentioned previously, data cannot be transported as part of a solution. This means an actual row within a table can't be included in a solution.

As an example, we can include the `Contact` table in a solution. This would include all the changed metadata such as columns, views, and forms. We cannot add a specific contact (for example, `Max Mustermann`) to a solution.

Some of the components developed by Microsoft are basically just rows in tables implemented by Microsoft. Still, we are able to transport them in solutions since Microsoft made them "solution-aware." Some examples of this type of component are duplication detection rules, environment variables, or connection references.

PPAC

Power Platform Admin Center (PPAC) is reachable at `https://admin.powerplatform.microsoft.com/`.

Maker portal

The **maker portal** is the website where one does all the customizations to a Power Platform solution. It is reachable at `https://make.powerapps.com/`.

Preview functionality

When Microsoft releases new features, it usually follows a certain path. First, the feature will only be released to a selected group of customers; this is called a **private preview**. After gathering feedback and improving the feature, Microsoft will release it in a public preview. This means that anyone can opt into using the functionality by activating the feature in the PPAC, installing some additional solution, or opting in through a different process. This depends on the feature in question. After a feature has passed the public preview, it normally becomes **General Availability (GA)**.

It is important to note that features that are in preview aren't meant to be used in a production implementation.

Release wave

Microsoft does release new functionalities in two **release waves** a year. The first one starts in April and the second one starts in October. They last for five months each. This means the first one ends in August and the second one in February. In this timespan, Microsoft releases new functionalities weekly with smaller deployments up to daily. The majority of new features are released in the first installation of each release wave and include all functionalities that are automatically activated for end-users.

This setup enables Microsoft to release many features throughout the year. It also gives you two months wherein no new features are released from Microsoft.

When exactly the first installation of every wave hits your environment depends on the region your environment is located in.

Release plans

Circa four months before every release, Microsoft releases the release plan to the wave. It contains all the features we would expect in a release wave. For every feature, Microsoft provides a detailed or somewhat detailed description, an estimated date when the preview starts, and an estimated date when the feature will be GA.

The **release plan** can change during a wave. This means that new features can be added, and existing features can be changed, deleted, or moved to a later time in the same wave or even a later wave.

Early access

Two months after the release plan is released and circa two months before the release wave is released, **early access** will be available. Early access can be activated for an environment and will install a subset of features of the upcoming release wave. Microsoft has promised to always include all features that impact end users directly in early access.

This is important so that you can test your current solution against the new release and fix any bugs before the release actually hits your production environment. We will learn more about this in *Chapter 3* when we talk about environment structure.

PAC

Power Apps CLI (PAC) uses a **Command Line Interface (CLI)**. In this case, a developer can easily run command line scripts to execute actions in Power Platform.

In a later version, it was renamed to Power Platform CLI. PAC, as the command when executed in the command line, is still the same. The reason is that Microsoft doesn't want to do a breaking change, as this would have meant that all scripts already created by customers would have stopped working.

We will learn more about PAC in *Chapter 5*.

Now that we have established an understanding of some of the important terminology when it comes to Power Platform, the next section will explain the difference between "normal" ALM and ALM in Power Platform.

"Normal" ALM versus Power Platform ALM

ALM in Power Platform is not fundamentally different from "normal" ALM, as described in *Chapter 1*, but in addition to what already has been described, there are some things to watch out for.

The main problem people have a hard time wrapping their heads around is the fact that in a "normal" implementation project, everything is done through some form of code, which is controlled through source control. In Power Platform, configuration and customization are done in an environment, which, on its own, isn't directly related to any form of source control.

> **Source code**
>
> There is a new feature in the current **Release 2024 Wave 2 Plan**, which describes a native Git integration to Azure DevOps. This is a promising feature that could solve parts of the problems we have when it comes to ALM in Power Platform.

In *Chapter 7*, we will learn an approach for establishing such a connection to source control. In *Chapter 10*, we will implement the approach described in *Chapter 7*.

Secondly, a solution in Power Platform often only contains a small part of classic development. The biggest part of Power Platform implementation is configuration or customization. It also is important to notice that the ratio shifts away from development more and more, with the huge investment by Microsoft into low-code tools in Power Platform.

In addition to the mentioned challenges and problems, the way Microsoft has implemented parts of the platform (mostly because of historical reasons) and some components can come with risks and challenges. For example, deploying connection references and environment variables isn't as trivial as one would expect. We will talk more about those in *Chapter 6*. Something that isn't included in this book, because it is too much for the scope of this book, is **service endpoints**. Those connect a Dataverse environment to an Azure Service Bus in a native way to ease creating integrations. Service endpoints, for example, add additional complexity to an ALM process if they are used, since there is a need for manual steps after every deployment. The mentioned manual steps can be automated using some custom scripts.

The next section will explain different solution levels and why those are important for ALM in the Power Platform.

Solution levels

While planning a solution, selecting the correct solution is crucial. It also tremendously affects the ALM process, which must be established. In this section, we will explain different solution levels. The idea is to always select the lowest level when possible and only go to the next level in case we can't implement a requirement with the current level. This means for every requirement we would start with the question "Can I implement this using Configuration?" if the answer is no we ask "Can I implement this using noncomplex Customization?" and so on.

Configuration

Configuration is, as mentioned, everything we can configure in the environment with out-of-the-box features without changing existing features or creating new ones.

This can be, for example, system settings or server-side sync.

Configuration is the first level and whenever a requirement can be implemented with "just" configuration, that should be the preferred way.

Noncomplex customization

The next level, for requirements that can't be implemented using configuration, is **noncomplex customization**. This can be, for example, a simple flow changing an existing view or form.

Complex customization

If a requirement isn't realizable with noncomplex customization, the next level is **complex customization**. Examples include complex flows, changes in the data schema (such as adding tables, fields, or relationships), or the creation of apps (model-driven or canvas).

Third-party product

In case none of the previous levels are feasible, we should consider using a third-party product if there is one on the market that can resolve the problem in question.

Depending on the maturity level of the team, as well as, for example, the price for the third-party product, this level could be substituted for the next level: development.

Development

Custom development should be done whenever none of the previous levels are possible. With custom development, we mean, for example, custom components developed with the **Power Apps Component Framework (PCF)**, plugins, custom actions, or web resources.

Preview features

Preview features are a bit of a gray zone. Microsoft does not recommend using them in production, but in some cases, it might be worth using them anyway and making the informed decision to take the risk that Microsoft might make fundamental changes to the feature or even retire it completely. This is mostly weighed against how much investment a custom development would be.

In particular, this level can add extra complexity to an ALM process since some preview features Microsoft releases aren't ready for a proper ALM process and need manual (or custom) scripts to work smoothly.

As you can see, choosing the correct level has implications for the ALM process. It can make the process more complex or the implementation of it easier.

Summary

In this chapter, we learned about important terminology when it comes to Power Platform.

We also talked about the difference between "normal" ALM and ALM in Power Platform. The biggest problem is to understand where which change is done and how it relates to a proper development process.

Lastly, we learned about different solution levels and why they are important for laying out a matching ALM process for a project.

The next chapter will go into detail about environments in Power Platform and why they are a crucial part of the ALM process.

Questions

1. What is the first solution level?

    A. Development

    B. Noncomplex customization

    C. Configuration

2. What describes the upcoming features for Power Platform?

    A. Release wave

    B. Release plan

    C. Early access

3. Which difference between "normal" ALM and Power Platform ALM is usually the hardest to understand?

    A. Most changes are not done in code

    B. A Power Platform solution only contains a small amount of custom code

    C. Challenges associated with how Microsoft has implemented some features

Further reading

Connect your environment to source control: https://learn.microsoft.com/en-us/power-platform/release-plan/2024wave1/power-apps/connect-environment-source-control

3
Power Platform Environments

This chapter will go into detail regarding Power Platform environments. It will explain what environments are, which types exist, as well as which strategies one could use.

In addition, we will talk about Landing Zones in Power Platform, as well as which settings there are for an environment. Some parts of this chapter will provide a short excursion into the field of governance.

We will cover the following main topics:

- Environment basics
- Environment types
- Environment strategies
- Tiers
- Landing Zones

By the end of this chapter, you will have discovered everything you need to know about environments in Power Platform.

Environment basics

Environments in Power Platform serve as a container for configurations, customization, and data. Here, you create, store, manage, and share the data and functionality of your business.

They can be used to separate apps, which have different security requirements or target audiences, as well as separate data in a secure way. This means that one can use a well-thought-through environment structure to make apps and related data more secure.

Each environment is created in a Microsoft Entra tenant and bound to a certain Power Platform region. No data or configuration will leave the selected region (with very few exceptions). All the apps, bots, flows, connections, gateways, and more are also created in the selected region.

> **Note**
> Every Power Platform region is built on top of at least two Azure regions. Every Azure region contains at least two data centers. It means a Power Platform environment is saved to four Azure data centers, as of the time of writing this book.

An environment can have zero or one Dataverse "databases." Dataverse provides a place to store data related to your app.

Environment types

There are several different types of environments, which are created automatically or can be created manually. In this section, we will explain the most important types.

Default

Every tenant has a default environment, which is named `{Microsoft Entra Tenant Name}` `(Default)`. The default environment is shared between all Power Apps users in that tenant. All new users will automatically get the Maker role. It also contains a Dataverse database.

The idea is to use this environment only for personal productivity apps.

> **Tip**
> An admin should consider changing the name of this environment to something such as `Personal Productivity`.

The default environment cannot be deleted and has a max capacity of 1 TB.

Development

Every user with the Developer Plan license, which can be obtained by anyone without any cost, is entitled to three environments of this type.

They can either be created by the user or the creation can be restricted to only admins.

The storage usage of development environments will not be included in the tenant's storage entitlement. However, they are limited to 2 GB of Database storage.

In addition, development environments can only contain Power Apps Dataverse and can't be used with first-party Dynamics 365 apps.

The intention was for development environments to only be used by the owner. Nowadays, other users of the same tenant can be invited to use development environments as well.

This type of environment will automatically be deleted when the owner is not actively using their Developer Plan license.

Since this environment is tied to the user and will automatically be deleted when the user is inactivated, the recommendation is to not use it as part of your application environment structure.

Sandbox and production

Sandbox and production environments are *full-blown* environments. One can use Dynamics 365 first-party apps as well as custom apps in them.

Access can be restricted through Entra ID security groups. The used storage is included as part of the tenant entitlement for Power Platform storage.

Usually, all environments of an environment structure are one of two types; that is, the production environment is of the type production and all other environments (such as development and pre-production) are of the type sandbox.

Trial

The last type worth mentioning is trial environments. They are used to test out new functionality or apps. They are not included in the tenant storage and will be automatically deleted after a certain amount of time.

Environment strategies

There are different strategies when it comes to how to set up your environments. Often, they are combined.

Minimal

The minimal setup includes the following:

- Development
- Test
- Production

As the names suggest, one would do all the implementation in the development environment, test new functionality in the test environment, and apply them in the production environment.

In the best-case scenario, developers only have access to development and test. Users only have access to production and super users/testers also have access to the test environment.

The following figure illustrates the explained minimal environment strategy.

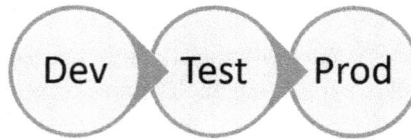

Figure 3.1: Minimal environment pipeline

In this section, we have learned what a minimal environment structure would look like. The next section explains another strategy.

Functional separation

One of the more common strategies is one that could be called **functional separation**. This is used when two different departments or branches in different countries should get different functionality and therefore different data in production. In that case, separate environment pipelines could be established since they don't share any parts.

The following figure illustrates this approach.

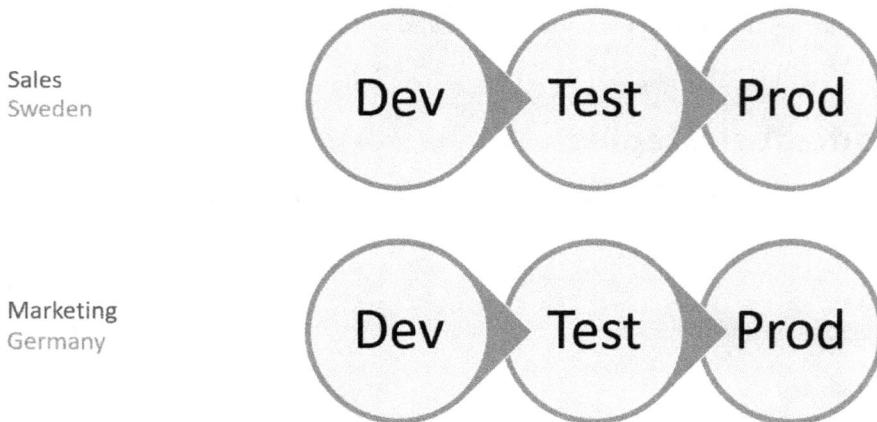

Figure 3.2: Functional separation

Data separation

In some cases, different departments or, more commonly, branches in different countries have the requirement to be separated on a data level, but the functionality they use should be the same.

In this setup, we have one development environment and separate test and production environments. Test and production both receive the same solution(s) or functionality.

Let's assume we have a company with branches in Sweden and Germany. One example could be that users from Sweden should under no circumstances be able to see data for users in Germany.

Another example could be that employees should never see data from the **Human Resources (HR)** department unless they work in HR.

The following figure illustrates this approach.

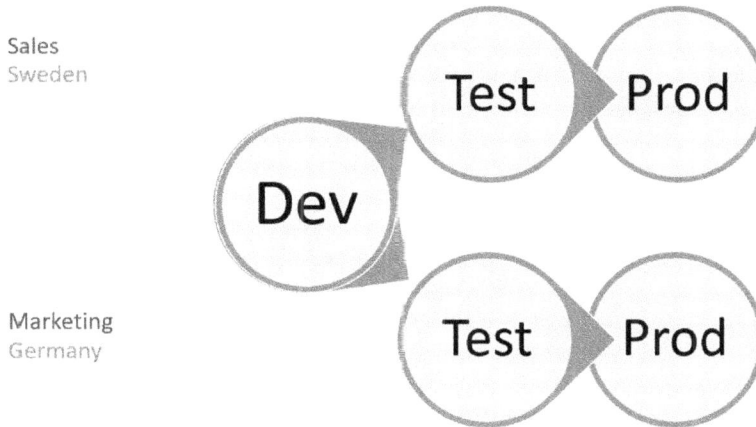

Figure 3.3: Data separation

Development separation

Another approach that is widely used is one called **development separation**. In this case, there are several development environments that aren't related to each other in any way. They all install their solution in the same test and production environments since the group of end users is the same.

This is often used for smaller apps that don't share any information and therefore could be developed separately.

The following figure illustrates this approach:

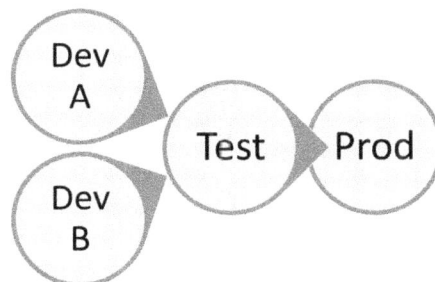

Figure 3.4: Development separation

Internal development

The last strategy is one I call **internal development**, mostly because we often use it when our customers would like to build on top of the implementation we provide. This can also be used when you have several development teams that all together build the product the end user is using. The end user usually doesn't even know that there are different development teams involved.

In this approach, the development environments are *placed next to each other* in the environment pipeline. That is, all solutions from prior development environments are installed and managed in the next development environment. A new solution is developed on top of the second development environment. All solutions together are then deployed to test and production.

The following figure illustrates this approach.

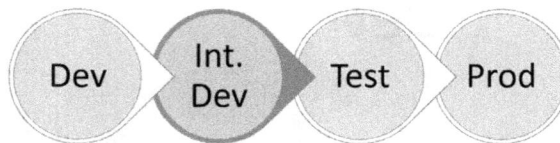

Figure 3.5: Internal development

Surrounding environments

In addition to the mentioned structure, a project could, depending on the requirements, have surrounding environments. Those can, for example, be the following:

- Evaluation
- Hotfix

Evaluation

As you might know, Microsoft releases new functionality in two waves, starting in April and October. Each wave stretches from the first deployment until around one month before the next wave starts. This means that in the April wave, new features will be released weekly from April until September.

> **Note**
> Depending on the region you are using, the first install of the wave will hit your environments at different times. Read more about the *Released versions of Microsoft Dataverse* here: https://learn.microsoft.com/en-us/dynamics365/released-versions/microsoft-dataverse.

Microsoft promises that all features affecting end users directly, without the possibility of an admin opting out, will be included in the first batch. They also promise that the same features are part of early access, which will be available around two months before the wave is released.

This early access should be used in your application with the new functionality, update documentation, and training, as well as train the end users on UI changes.

Since the new functionality in the wave and the early access can break your deployment to downstream environments, these tests should be done in a separate environment: the evaluation environment.

Hotfix

To be able to fix issues that are detected in production, a hotfix environment is important. Usually, when a bug is detected, the development environment already contains changes that should not be deployed to production and therefore the development environment isn't suitable for fixing those bugs.

Solutions in a hotfix environment should, like the development environment, be unmanaged, whereas in all other environments, they should be managed. In the next chapter, we will learn what this means exactly.

Tiers

When it comes to governance and when to implement a robust ALM process, Microsoft talks about three different tiers of applications in Power Platform.

Productivity

Apps in this tier are mostly for personal productivity only. They are usually built by citizen developers and don't need a high level of governance. A high iteration rate and missing review or approval process are typical for those apps.

Apps in this tier do not have or need an ALM process.

Important

When the app gets more complex, more important, or higher in scope (more users using it), it moves to the next tier, which is important. Apps in this tier should be known and approved by IT.

A proper ALM process might be applicable but is not necessary. Often, Power Platform pipelines are used for this kind of application.

Mission critical

The last tier, according to Microsoft, is mission critical. All apps that are mission critical, used across the organization, or have high complexity or high risk (use of personal data and therefore GDPR risk, for example) should be in this tier.

An app that is moved into this tier is usually redeveloped by IT or professionals.

All apps in this tier should have a robust ALM process setup.

> **Note**
>
> The exact criteria of when an app is "too" complex or mission critical differ from company to company.

Landing Zones

Microsoft has created a reference implementation and architecture and design methodology for Power Platform called Landing Zones.

It allows you to scale the use of Power Platform and automate all the "plumbing" that is needed when an app is to be created. One can create "templates" for different types of setups to easily deploy new environment pipelines into the tenant when needed.

It considers all the required platform parts, which include the following:

- Access management
- Governance and compliance
- Connectivity, interoperability, and extensibility

Different parts of the architecture are for different roles, such as platform operations, security operations, and network operations, as well as design principles such as policy-driven governance, persona agnostic, and a single control and management plan.

The reference implementation includes several different governance structures, such as the following:

- Tenant governance
- Admin environment governance
- Default environment governance
- Citizen developer Landing Zones
- Professional developer Landing Zones

As you can see, Landing Zones are a huge topic that includes both specific implementation (as the reference implementation) and a lot of guidance and best practices.

In my experience, they are most useful to big companies with a lot of users. For small to medium companies, however, it is usually too much overhead.

As mentioned, Landing Zones are a very big topic and are beyond the scope of this book to cover in detail. I have therefore provided some additional links in the *Further reading* section for you to explore further.

Summary

In this chapter, we have learned what environments in Power Platform are and what the different approaches to structuring them are.

We also covered the different tiers Microsoft uses to determine the amount of governance and ALM an app needs. We then briefly focused on what Landing Zones are, what they include, and when to use them.

All of this is important to be able to set the correct environment strategy for your project.

In the next chapter, we will talk in detail about solutions and what we can do with them.

Questions

1. Which environments are needed for the minimum strategy?

    A. Development, test, production

    B. Development, test, staging, production

    C. Test, production

2. Which environment type is not included in the tenant space entitlements?

    A. Sandbox

    B. Production

    C. Development

3. Which is the highest tier of apps?

    A. Productivity

    B. Mission critical

    C. Important

Further reading

Read more about Landing Zones:

*   `https://github.com/microsoft/industry/blob/main/foundations/powerPlatform/referenceImplementation/readme.md`
*   `https://github.com/microsoft/industry/tree/main/foundations/powerPlatform`
*   `https://www.microsoft.com/en-us/power-platform/blog/2022/02/18/north-star-architecture-and-landing-zones-for-power-platform/`

4

Dataverse Solutions

In this chapter, we will deep dive into Dataverse Solutions. We will cover everything you need to know about them. Besides the basics, such as managed versus unmanaged, update versus upgrade, and solution layering, we will discuss solution segmentation, how to work with solutions, and how to export and import solutions.

We will cover the following main topics:

- Basics of Dataverse Solutions
- Solution segmentation
- Working with solutions
- Exporting and importing

By the end of the chapter, you will have learned what Dataverse Solutions are, how to use them in your daily work, and some ways to segment solutions if needed.

Basics of Dataverse Solutions

Let's begin with the very basics of what solutions are in Dataverse. Essentially, they are containers. A Solution can contain certain customizations for the Power Platform and is used to transport those customizations from one environment to another.

When two unmanaged solutions contain the same component, a change to this component in one solution will be reflected in the second solution as well. This is because solutions basically only contain pointers to components, which means they don't contain the component or customization it represents.

Customization versus data

As mentioned earlier, solutions only can contain customization and no data. This section will explain the difference.

Customization is everything that can be done in the maker portal, or all the metadata to define a solution. This includes the following:

- Tables:
    - Fields
    - Relationships
    - Views
    - Forms
- Business rules
- Power Automate flows (both Cloud flows and Desktop Flows)
- Canvas apps
- Model-driven apps
- Custom pages
- Dashboards
- Security roles
- Choices

Data means specific rows in a table, for example, a representation of a contact in the contact table.

In addition to either customization or data, there are certain components that are only rows in a table, but Microsoft has made them "solution-aware." This means that even though they only are data, it is possible to include them in a solution. Some examples are as follows:

- Connection references
- Environment variables
- Settings
- Duplication detection rules
- **Automatic Record Creation (ARC)** rules
- **Service-Level Agreements (SLAs)**

In *Chapter 6*, we will learn in detail how connection references and environment variables work.

Managed versus unmanaged

There are two different ways that a solution can be represented within an environment: unmanaged or managed.

Unmanaged solutions are used for development. Components can be added to and removed from unmanaged solutions. Unmanaged components can be deleted from the environment. An unmanaged solution can be exported either as unmanaged or managed. When an unmanaged solution is deleted from an environment, all components that were in the solution will remain in the environment.

Managed solutions, on the other hand, can't be changed or exported. This means that you cannot add or remove components to or from a managed solution unless an update (or upgrade) of the solution is installed. When a managed solution is deleted from an environment, all components that are included in the solution will be deleted as well.

> **Exception**
>
> There is one exception to that rule. If the same component is included in another managed solution of the same publisher, it will not be deleted from the environment.

It is good to know that if a table or field is deleted the underlying data is permanently removed as well.

In addition, managed solutions can be used to clean up components that have been deleted from the development environment.

In all environments that are not development environments, managed solutions should be used. The only exception is a potential hotfix environment.

Update versus upgrade

When it comes to importing a solution into an environment there are two different import methods, update and upgrade.

Updates can be performed for both unmanaged and managed solutions. If a solution is installed using an update, new components will be installed and existing components will be updated.

An **upgrade** installation will also update existing components and install new ones. In addition, it will delete components that aren't a part of the solution any longer.

> **Cleanup**
>
> Since an upgrade will delete components that aren't included in the solution any longer, it is a very effective way of cleaning up a downstream environment without needing to delete components manually.

There are some requirements for performing an upgrade on a managed solution:

- The solution is already installed in the target environment
- The target version is higher than the version that is already installed

---

Speed up the import

Microsoft has invested a lot of energy in the whole ALM story. One part that has been improved a lot is the import speed with certain configurations.

An update with overwrite unmanaged customizations and convert to managed set to false, and an in-place upgrade (Stage and Upgrade) are the two fastest ways to import a solution.

We will take a deeper look at those settings in *Chapter 5* and *Chapter 10*.

Solution layering

In Power Platform, we have the concept of **solution layering** to determine what the end user is seeing.

Every installed managed solution, including the system solutions, gets a new layer. The order of the layers is determined by the date the solution was installed for the first time; newer solutions stack on top of older solutions. This means that the layer will not change when a new version of one of the solutions is installed.

On top of all managed layers, there is one unmanaged layer where all unmanaged customizations reside.

The layering is one of the reasons why you should use managed solutions in downstream environments. If solutions are installed as unmanaged solutions, the solution that was installed latest will always win. This means you always have to install all solutions in the exact same order. When a managed solution is used, a solution only has to be installed if it has changed.

Different layers can change the same component. In this case, a higher layer will "win" over a lower layer. The combination of all of those layers is what the end user will see.

The following image illustrates what we have discussed:

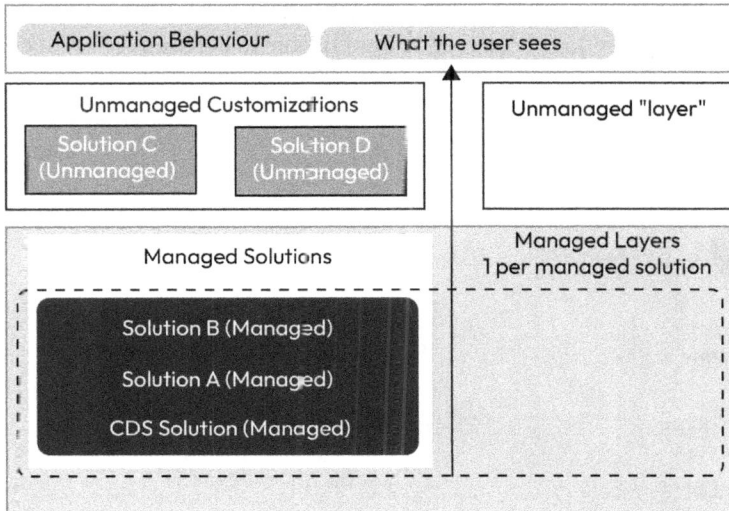

Figure 4.1: Solution layering

> **Layering problems**
>
> When Microsoft or an **Independent Solution Vendor** (ISV) is introducing a new solution, it will be layed on top of your solution. This could mean that the new solution wins over your customization.
>
> To fix this issue when it happens, you would have to create a new solution, move your components over, install it in the downstream environment, and delete the old solution.

Publisher

Every solution has a publisher. The publisher defines for example the prefix of new components as well as the start ID for new choices.

The publisher of a solution is considered the owner of the components in the solution. They can define what can be done with components.

As mentioned earlier, it is possible to move components between solutions with the same publisher. A component that has been installed with a certain publisher needs to be a part of at least one solution of that publisher.

This information is important when it comes to solution segmentation, and especially restructuring solutions.

Now that we have learned the basics when it comes to solutions, we will take a look at Patch and Clone in the next section.

Patch and clone

There are two types of solution versioning: patch and clone. They were created as a workaround when the import speed was very poor. With patches, it was possible to deploy a subset of components to increase import speed.

These options are still there in the maker portal:

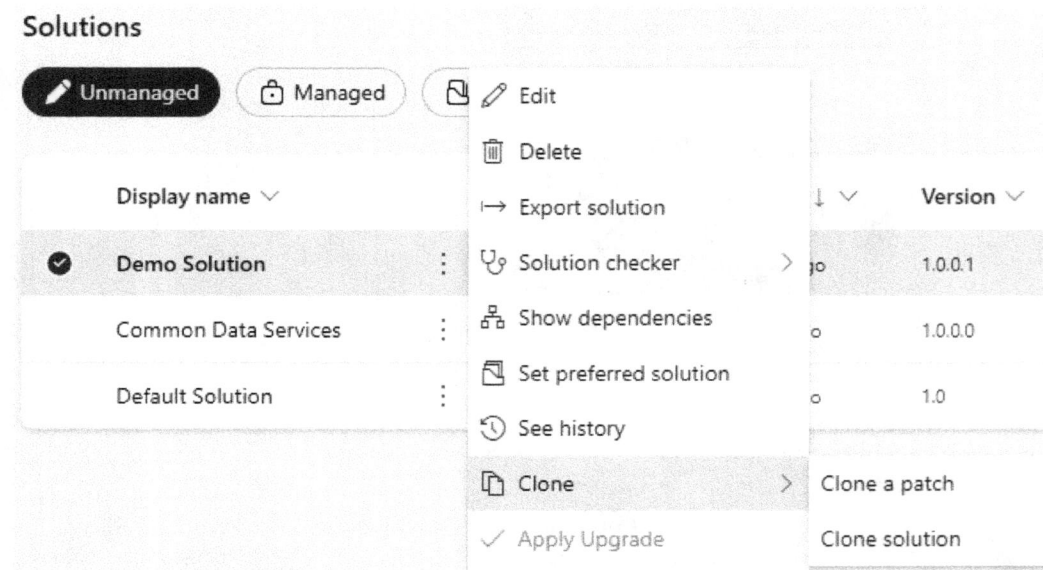

Figure 4.2: Patching and cloning solution

The idea is that a clone will change the first two parts of a solution version. It will also be a copy of the base solution and contain all components.

A patch will only change the last two parts of the version and will be empty. We can then add components to it and move them to a downstream environment.

All patches will be rolled up into the solution when the base solution is cloned.

The recommendation is to not use patches. They were created as a workaround and are not recommended by Microsoft any longer. In addition, patches make an automated ALM process nearly impossible. One reason is that the solution name changes with every patch, and when a patch is installed into a downstream environment the base solution present has to be the same version as the base solution the patch was created on.

Patches are good for hotfixes, though.

Solution segmentation

In this section, we will look at how to segment solutions when you need to have several solutions.

The general recommendation is to have as few solutions as possible, mainly because having several unmanaged solutions sharing the same components will create cross-dependencies, which might be hard to resolve in the future.

Historically, one reason to segment components in different solutions was poor import speed. In that case, it would have been possible to only deploy parts of the implementation by only deploying a subset of solutions. Since Microsoft has increased the import speed tremendously, this isn't really a valid point any longer.

> **Import speed**
>
> In the last section, we mentioned the configuration to use the improved import speed. I would like to give you an example of how big the improvement has been. For one customer, we use a mono-solution approach (we have two extra solutions for certain components). The solution contains around 450 components. Previously, the import took around 70-80 minutes. Using the right configuration, we can bring down the import to 4-5 minutes.

There are still several reasons why segmented solutions might be needed:

- Team structure
- Different suppliers/vendors
- Partial releases
- Use of certain components

As mentioned, certain components require a separate solution. They are as follows:

- SLAs
- ARCs
- Custom connectors

If the project is using those it is important to have separate solutions for them. Otherwise, if they are deployed in a solution with all other components the import might fail.

> **One environment per solution**
>
> A recommendation from Microsoft is to only have one unmanaged solution per environment. If the implementation needs solution segmentation, every solution should get its own environment. In my opinion, that is, for most projects, not doable.

If it is really necessary to segment project solutions, there are basically two different approaches to choose from, vertical and horizontal segmentation.

Vertical segmentation

Let's take a look at vertical segmentation first. Here, we create one solution per area. This could be, for example, different departments in a company. All components the area uses are included in that one solution.

The following screenshot illustrates this approach:

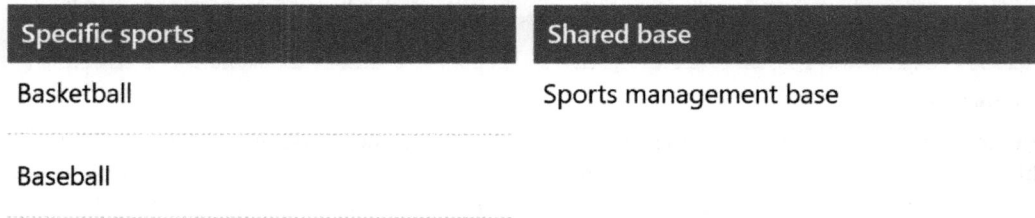

Specific sports	Shared base
Basketball	Sports management base
Baseball	

Figure 4.3: Vertical segmentation

With this approach, it is easy to understand which area is using which components.

Vertical segmentation will most likely lead to cross-dependencies very quickly.

Horizontal segmentation

The second approach is horizontal segmentation. In this approach, there is one solution per component type. All components of that type, no matter which part of the company they belong to, will be included in the respective solution.

The following screenshot illustrates this:

Visual Components	Processes & Plugins	Reports	Security Roles	Main
Apps	Processes	Reports	Security roles	Tables
Canvas components	Plug-in assemblies			Choices
PCF components	SDK message processing steps			Client extensions
Web resources				Service endpoints
				Dashboards
				Connection roles
				Email templates
				Column security profiles

Figure 4.4: Horizontal segmentation

This approach will minimize dependencies, not erase them completely, and allow a partial deployment.

However, it decreases transparency with regard to which part of the company is using which components.

In my opinion, horizontal segmentation is the only type of segmentation that is manageable.

Supplier separation

Another important part to mention when it comes to solution segmentation is **supplier separation**.

This basically means that it is important to have a structure for separating different suppliers. They could be different external companies or internal teams working together on the implementation.

In general, every team or supplier should have its own environment. Here, all the solutions of the previous teams/suppliers would be installed as managed solutions.

It is important to note the publisher in this setup. All solutions from one environment/team should use the same publisher. Different teams should usually have different publishers.

In the next section, we will learn how to work with solutions. This includes creating solutions and adding components.

Working with solutions

In this section, we will take a look at how to work with solutions.

Creating a solution

First of all, we need to know how to create a solution.

In the maker portal (`https://make.powerapps.com`), go to **Solutions**. There, click on + **New solution**.

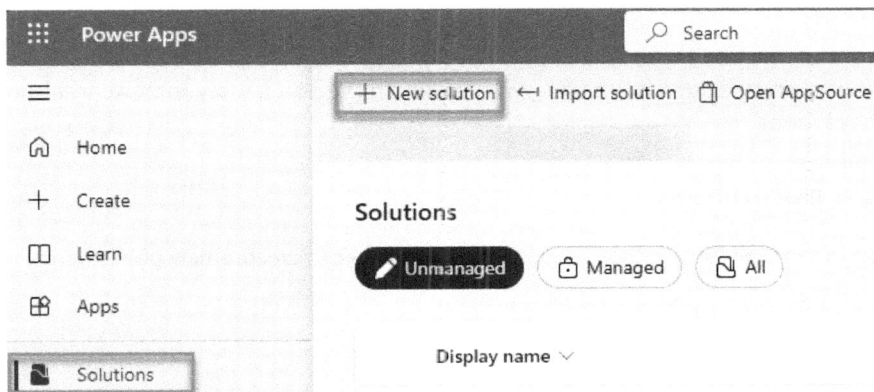

4.5: Opening the New solution flyout

This will open the form for creating a new solution on the right side. Here, we have to present a **Display name** and a **Name** (a schema name without spaces). We also have to specify the publisher by selecting one from the list or creating a new one.

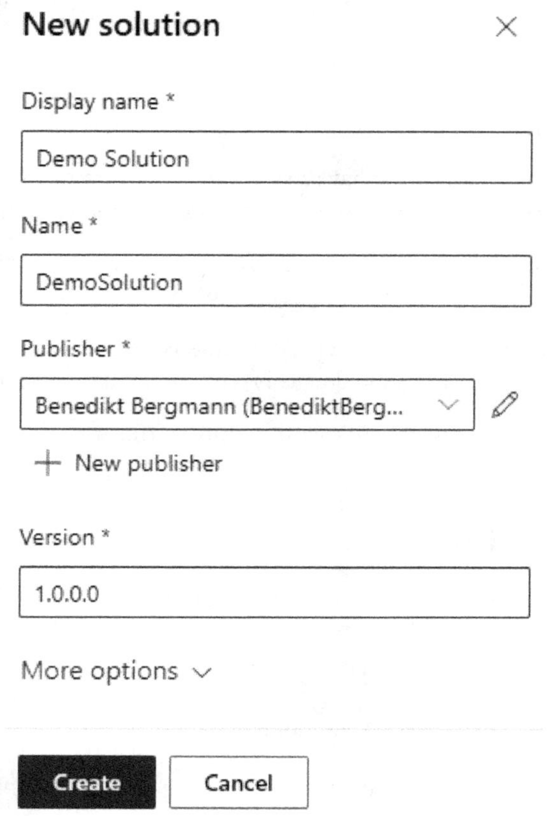

New solution   ✕

Display name *

    Demo Solution

Name *

    DemoSolution

Publisher *

    Benedikt Bergmann (BenediktBerg... ⌄   🖉

    ＋ New publisher

Version *

    1.0.0.0

More options ⌄

    Create   Cancel

Figure 4.6: Creating a new solution

With this, we have created our first unmanaged solution. In the next few sections, we will learn how to work with solutions.

Creating a publisher

While creating a new solution, you can select + **New publisher** to create a new publisher.

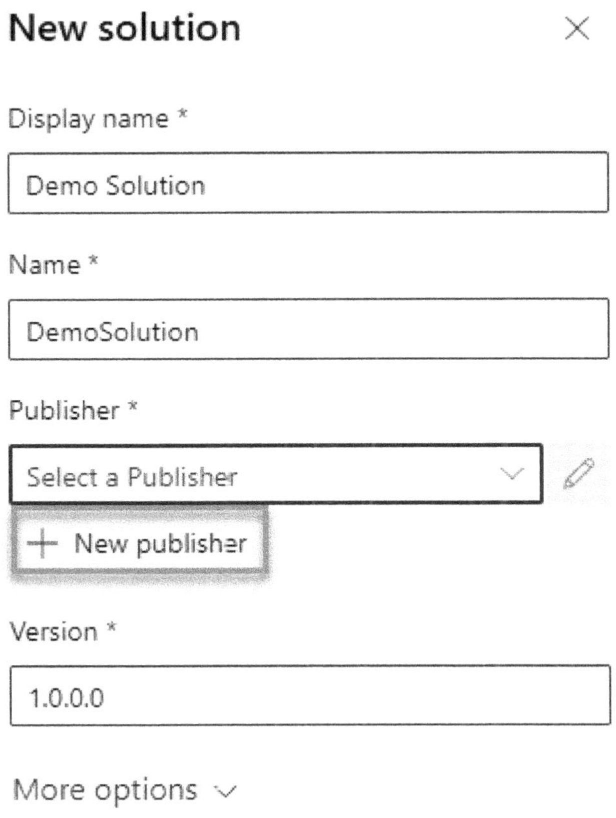

New solution   ✕

Display name *

Demo Solution

Name *

DemoSolution

Publisher *

Select a Publisher   ⌄   ✐

＋ New publisher

Version *

1.0.0.0

More options ⌄

Figure 4.7: Opening the New publisher flyout

This will open another flyout that contains the form for a new publisher. In this form, you can specify the following information:

- **Display name**
- **Name** (the schema name without spaces)
- **Description**
- **Prefix**
- **Choice value prefix** (this will be autogenerated based on the prefix)

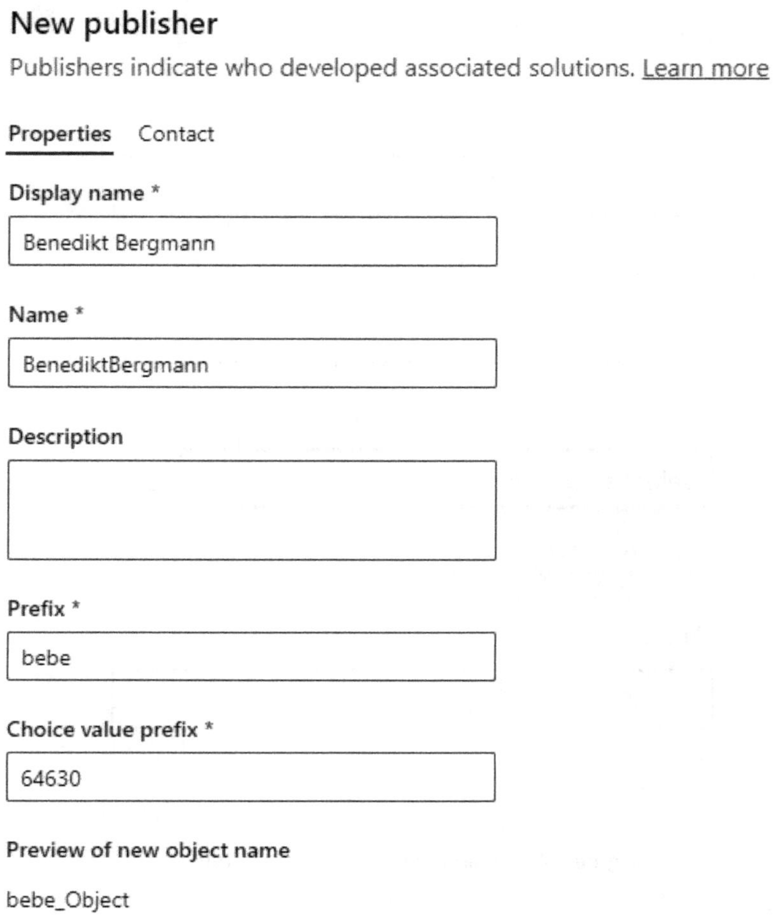

New publisher

Publishers indicate who developed associated solutions. Learn more

Properties   Contact

Display name *

> Benedikt Bergmann

Name *

> BenediktBergmann

Description

>

Prefix *

> bebe

Choice value prefix *

> 64630

Preview of new object name

bebe_Object

Figure 4.8: New publisher

In addition, you can add more contact information on the **Contact** tab of the creation form.

Navigating in a solution

If we open a solution, on the left is a menu containing **Overview**, **Objects**, **History**, and **Pipelines**.

In **Overview**, we have **Details** about the solution, such as **Name**, **Created on**, **Version**, and much more. In addition, we can see an overview of the solution status, including when the solution checker was run the last time, an overview of the Dataverse search indexes, and a list of recent items.

In addition, there is a top navigation (also called ribbon) where we have stuff such as **Publish all customizations** and **Export**.

The following screenshot shows the side menu, the **Details** box, and the ribbon.

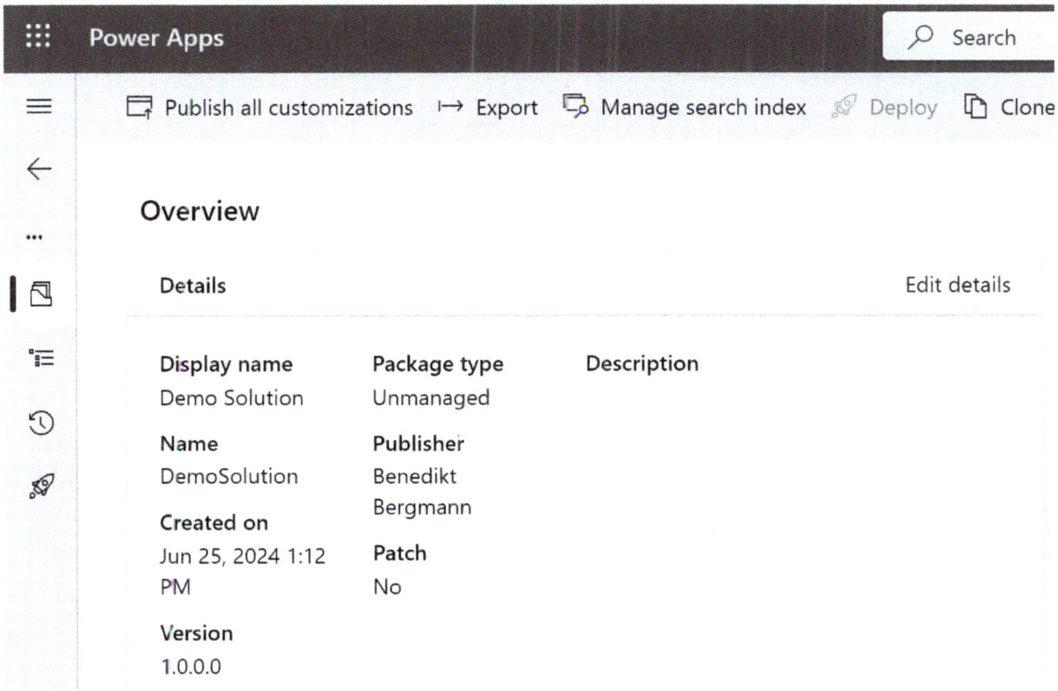

Figure 4.9: Solution Overview – Part 1

The next screenshot shows the solution status and the Dataverse search.

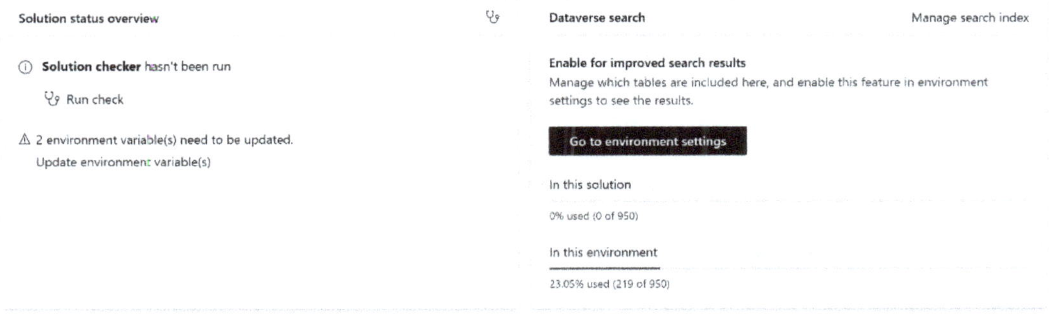

Figure 4.10: Solution Overview – Part 2

The next screenshot shows the list of recent items.

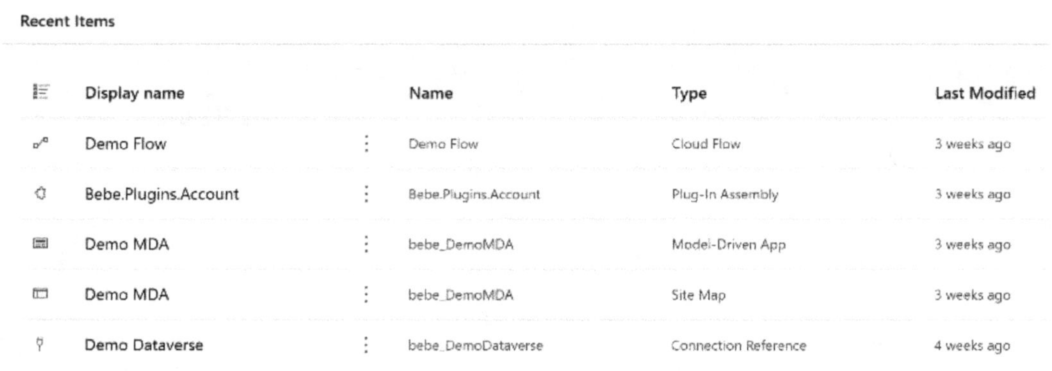
Recent Items
Display name
Name
Type
Last Modified
Demo Flow
Demo Flow
Cloud Flow
3 weeks ago
Bebe.Plugins.Account
Bebe.Plugins.Account
Plug-In Assembly
3 weeks ago
Demo MDA
bebe_DemoMDA
Model-Driven App
3 weeks ago
Demo MDA
bebe_DemoMDA
Site Map
3 weeks ago
Demo Dataverse
bebe_DemoDataverse
Connection Reference
4 weeks ago

Figure 4.11: Solution Overview – Part 3

Objects contains another sub-menu of all the component/object types included in the Solution, as well as a list of all the components. This page is the default one that will be shown when a solution is opened.

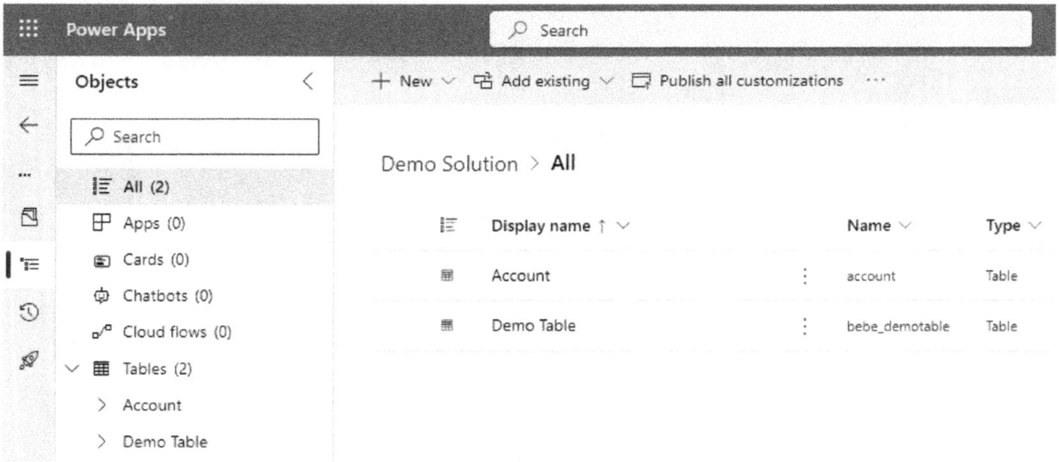
Power Apps
Search
Objects
+ New
Add existing
Publish all customizations
Search
All (2)
Apps (0)
Cards (0)
Chatbots (0)
Cloud flows (0)
Tables (2)
Account
Demo Table
Demo Solution > All
Display name
Name
Type
Account
account
Table
Demo Table
bebe_demotable
Table

Figure 4.12: Solution objects

In the left sub-menu, where all the components are shown, we can directly go to any component to see its details.

The **History** page shows a list of exports and imports of the solution.

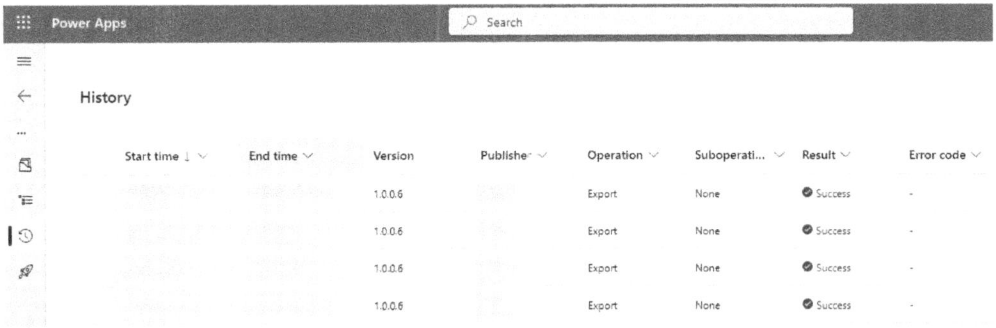

Figure 4.13: Solution history

The **Pipeline** option shows all related Power Platform pipelines. In *Chapter 9* and *Chapter 10*, we will learn more about Power Platform pipelines.

Adding components

Next, we have to add some components to our new solution. To do this, we navigate to the object page within a solution. We can either add new or existing components.

Add new components

Let's begin with the easier part of adding new components. On the object page in a solution, we can select + **New**, which is a drop-down menu. It offers all kinds of different components that we can create.

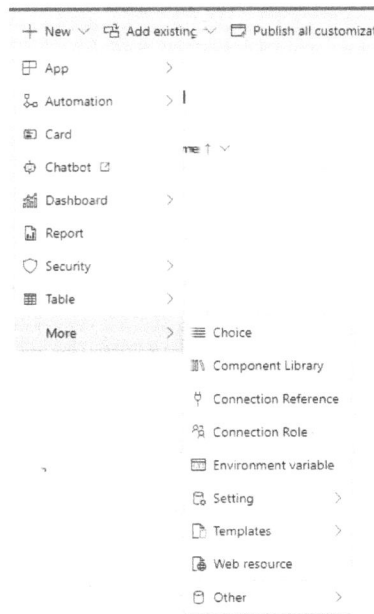

Figure 4.14: Creating a new component

By selecting one of those components, the respective editing experience will be started. The created component will automatically have the publisher's prefix, where it applies.

Add existing components

In the same way, we also can add existing components by using the **Add existing** button instead.

This will open a flyout from the right with a list of all components of the selected type. In this list, we can select one or more components. For most component types, we can click **Add** after selection.

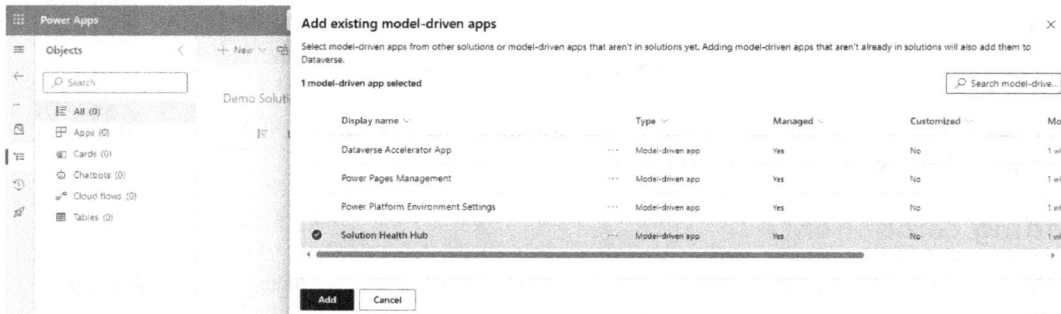

Figure 4.15: Adding an existing component

However, for tables, we will come to a second page after clicking **Next** when we make our selection.

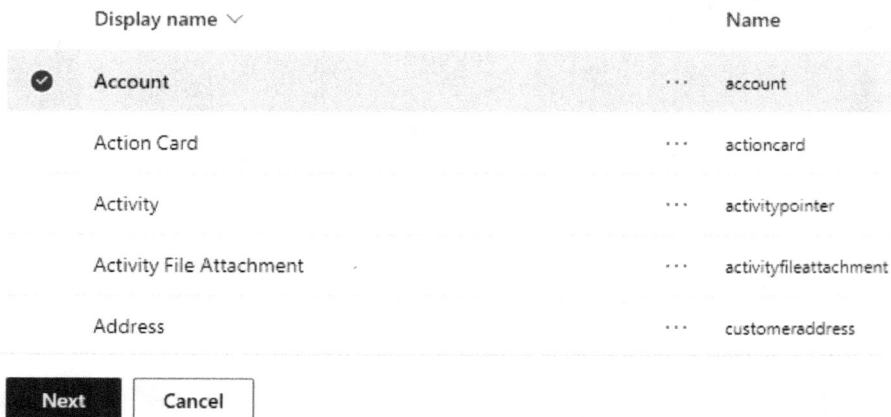

Figure 4.16: Adding a table

On the second screen, we can select specific objects from a table, as well as **Include all objects** and **Include table metadata**. **Include all objects** will, as the name suggests, include all the objects in a table. **Include table metadata** will include configuration at the table level.

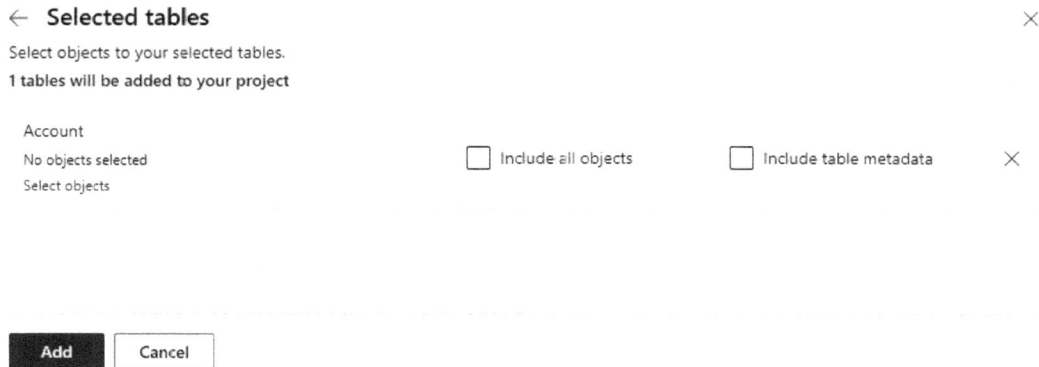

← **Selected tables**   ✕

Select objects to your selected tables.
1 tables will be added to your project

Account
No objects selected   ☐ Include all objects   ☐ Include table metadata   ✕
Select objects

Add   Cancel

Figure 4.17: Include all objects and Include table metadata

Include all objects should only be used for custom tables.

> **Never Include all**
>
> Tables that aren't from you, so either from Microsoft or the ISV, should never be added to your solution using **Include all objects**. This could lead to dependency issues that are very complex to resolve in the future.

Either way, we should try to only add objects from a table that we really need.

Removing components

Sometimes it is necessary to remove components from a solution when it is not used any longer. To do this, we navigate to the object list of our solution. Here, we find two possible ways of removing components.

The first way is through the three vertical dots, which open a context menu. There, we can select **Remove** and then either **Remove from this solution** or **Delete from this environment**.

Demo Solution > **All**

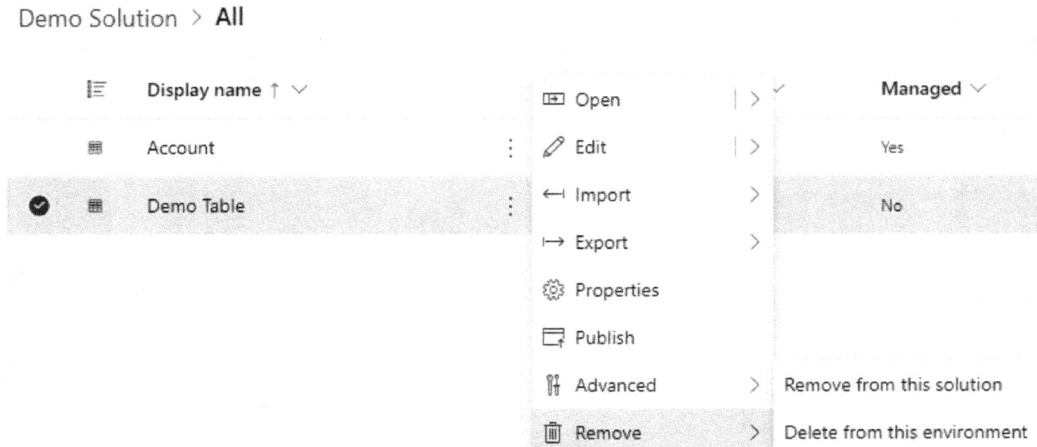

Figure 4.18: Deleting a component using the context menu

The second way is by selecting one or more components and using the **Remove** button on the ribbon navigation.

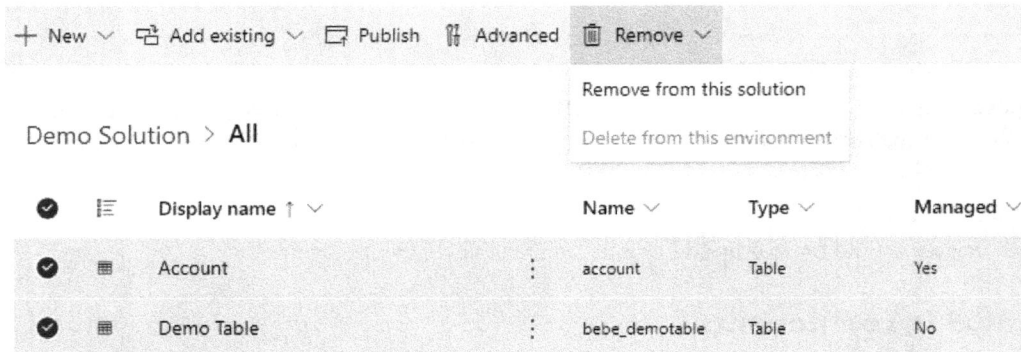

Figure 4.19: Removing a component using the ribbon

Remove from this solution will, as the name suggests, remove the component from the current solution. It will still remain in the environment. This is available for all types of components, both unmanaged and managed.

Delete from this environment will delete the component completely from the environment. It will therefore also be deleted from all other unmanaged solutions containing it. This option is only available for unmanaged components.

Set preferred solution

Every change made in an environment should always be done from within a custom solution. **Default Solution** and **Common Data Service Default Solution** should never be used.

To make this easier, Microsoft has introduced a new part of the platform, which is called **preferred solution**. In every environment, a maker can mark an unmanaged solution as preferred. All changes they make outside of a solution will be added to the preferred solution. In addition, the publisher of the preferred solution will be used for all those changes as well.

If a user has not selected a solution as **Preferred solution**, **Common Data Service Default Solution** is selected as the preferred solution by default. To change that, navigate to the list of solutions in the maker portal. There are two different ways of changing the preferred solution.

At the top, there is a button called **Set preferred solution**, which is always available.

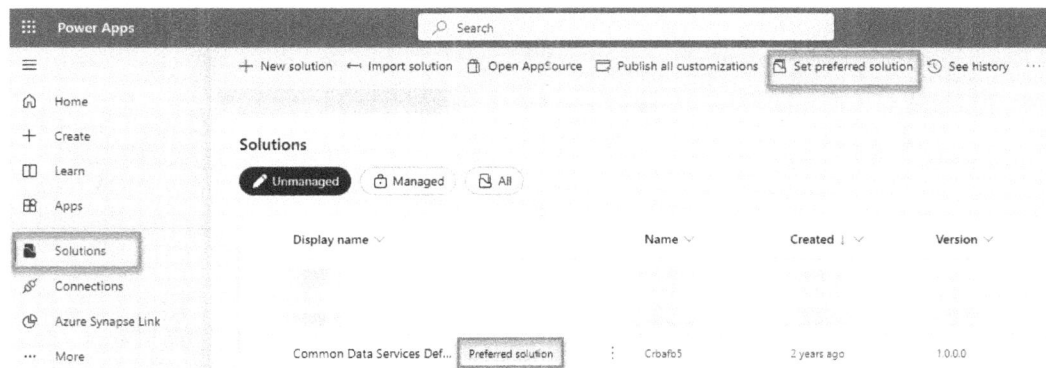

Figure 4.20: Opening the preferred solution dialog

Or there might be a big banner showing which solution currently is set as the preferred solution. There is a **Manage** button here.

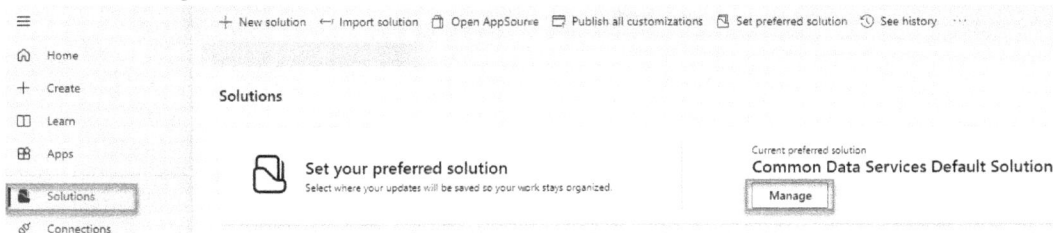

Figure 4.21: Preferred solution banner

Both open the same dialog to set the preferred solution. In the dialog, a user can select or create an unmanaged solution in the environment. The checkbox at the bottom will show the mentioned banner again.

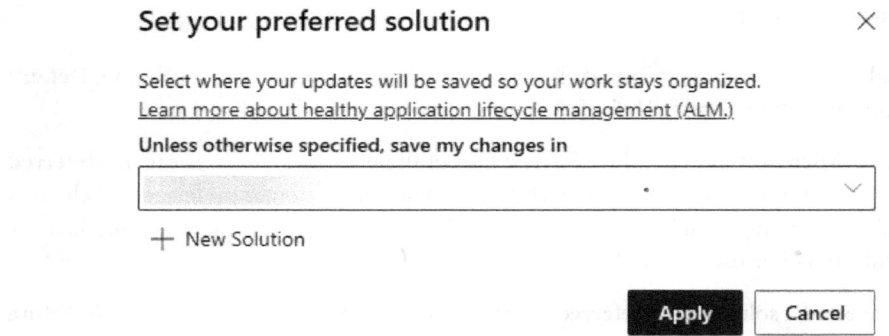

Figure 4.22: Set preferred solution dialog

Now that we have learned how to set the preferred solution to make sure our changes always end up in a solution, in the next section, we will learn how to export and import solutions.

Exporting and importing solutions

An essential part of every ALM process is to export and import solutions. The idea is that we can export a solution containing our customizations from a development environment and import it to a downstream environment. We will learn more about the specifics of the different approaches in the coming chapters.

Export

First, let's take a look at how to export a solution. To do that, we go to the list of solutions in the maker portal.

Here, we have two ways of exporting a solution, the same as we had for removing components from a solution.

We can use the context menu in a solution by clicking the three vertical dots beside the name. In that context menu, we can select **Export solution**.

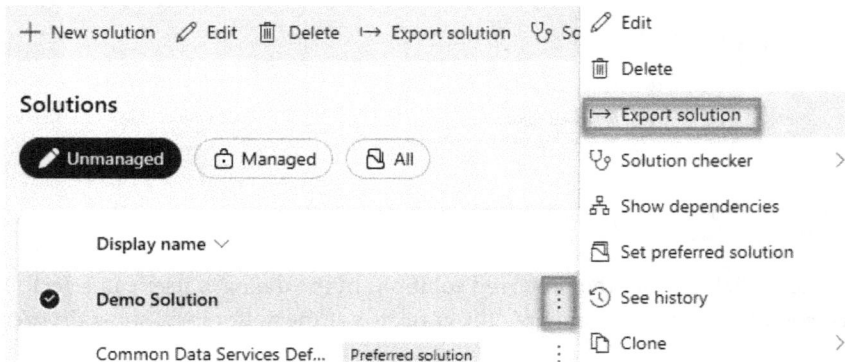

Figure 4.23: Exporting a solution via the context menu

The second approach is to select a solution from the list and use **Export solution** from the ribbon menu.

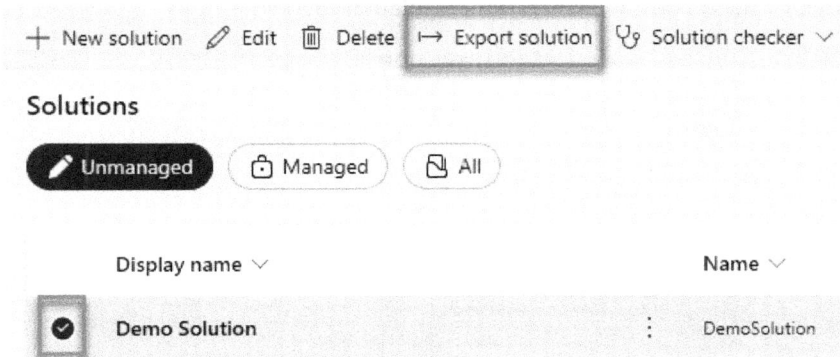

Figure 4.24: Exporting a solution via the ribbon

Both approaches will open the export wizard in a side pane on the right.

On the first page, we will have the option to publish all customizations and get some information about Power Platform pipelines, which we will learn more about in *Chapters 8*, *9*, and *10*.

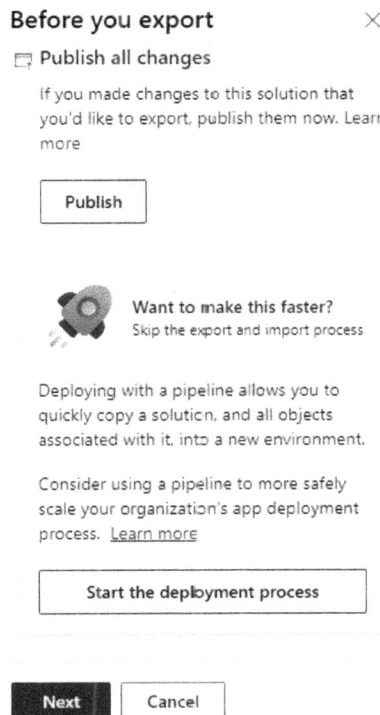

Figure 4.25: Publishing customizations

On the following screen, we can set the version to export, whether it should be a managed or unmanaged export, and whether the export should first run the solution checker.

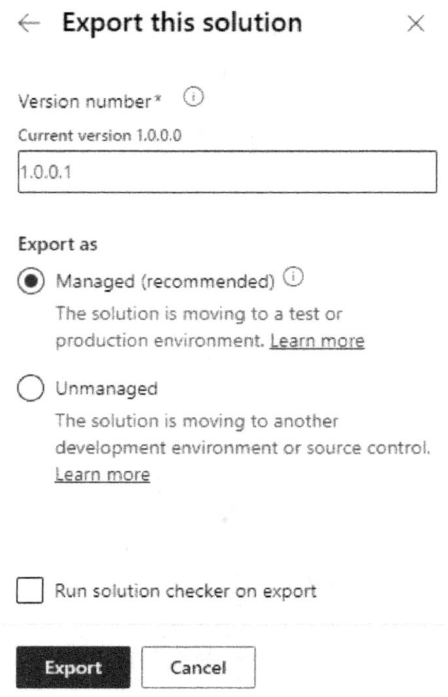

Figure 4.26: Exporting a solution

When we press the **Export** button, the side pane will close, and we will get a notification that our solution is being exported.

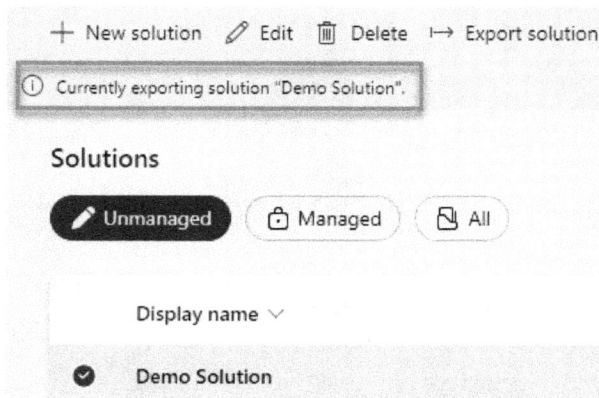

Figure 4.27: Export ongoing

When the export is ready, which will take longer for larger solutions, the notification will change to a success notification with a **Download** button.

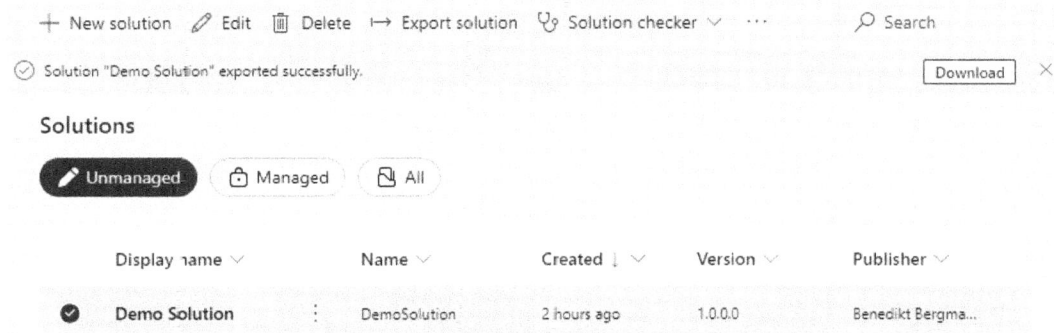

Figure 4.28: Downloading a solution

The exported solution can then be used to execute an import in a downstream environment.

Later, we will learn about automated ways to export a solution.

Import

To import a solution, we navigate again to the solution list in the maker portal. This time, we use the **Import solution** button in the ribbon.

This will open the import wizard in a side pane on the right side. Here, we can select the solution zip file to import by clicking the **Browse** button.

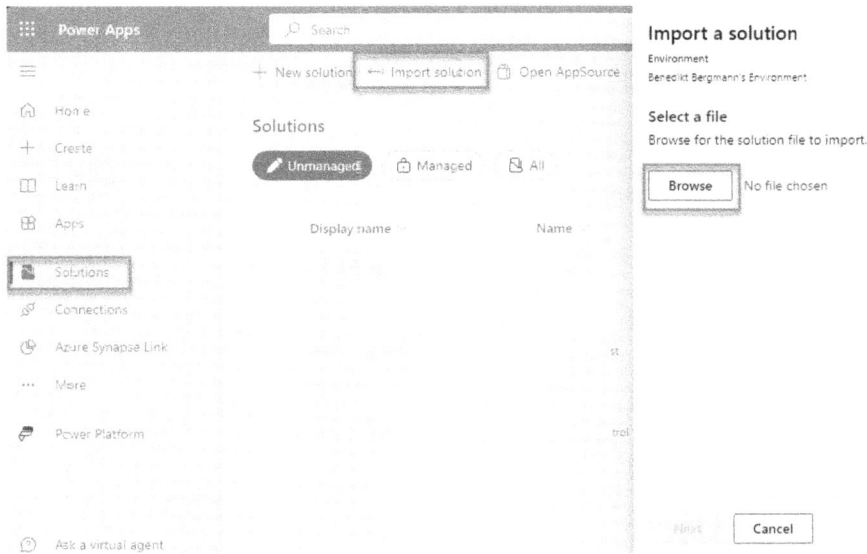

Figure 4.29: Import wizard page 1

After we have selected a solution zip file, the name will be shown beside the **Browse** button.

After clicking **Next**, we will come to the details page, where we can see more information about the solution as well as the current environment directly under the title.

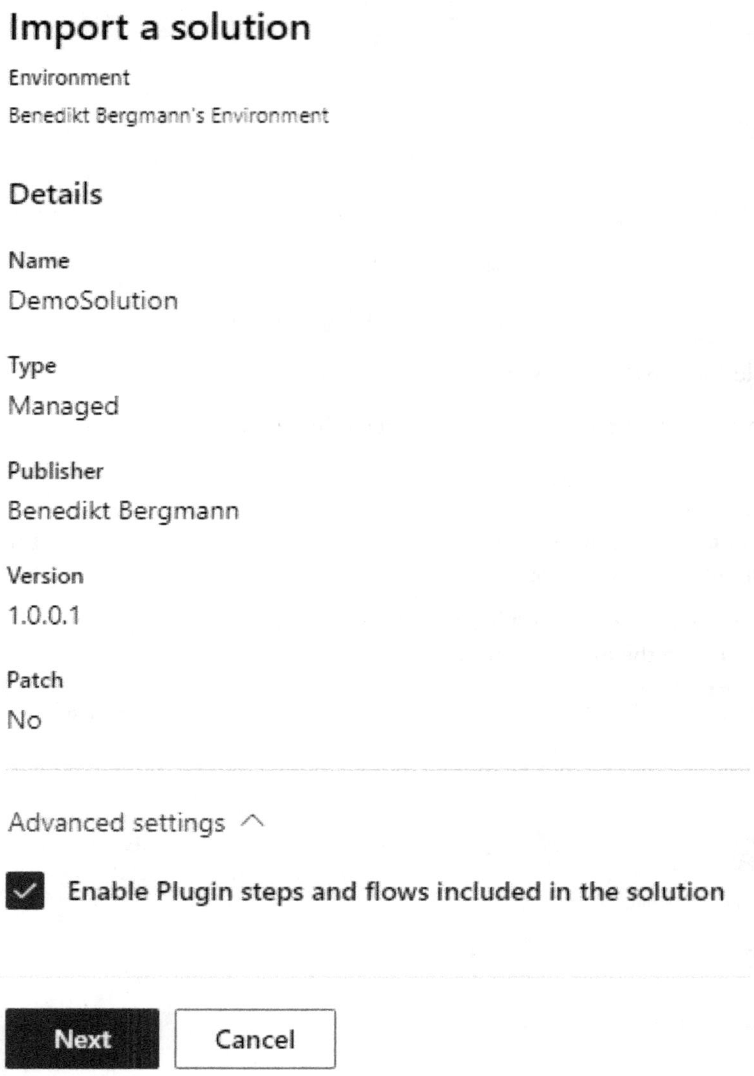

Import a solution

Environment
Benedikt Bergmann's Environment

Details

Name
DemoSolution

Type
Managed

Publisher
Benedikt Bergmann

Version
1.0.0.1

Patch
No

Advanced settings ∧

✓ **Enable Plugin steps and flows included in the solution**

Next   Cancel

Figure 4.30: Import Wizard page 2

On this screen, we either get a **Next** or **Import** button. It depends on whether you have connection references or environment variables in your solution. We will learn more about them in *Chapter 6*. In our demo solution, we have both, so if we click the **Next** button, we will come to the third screen of the wizard, which lets us map connections in the target environment to the connection references in our solution.

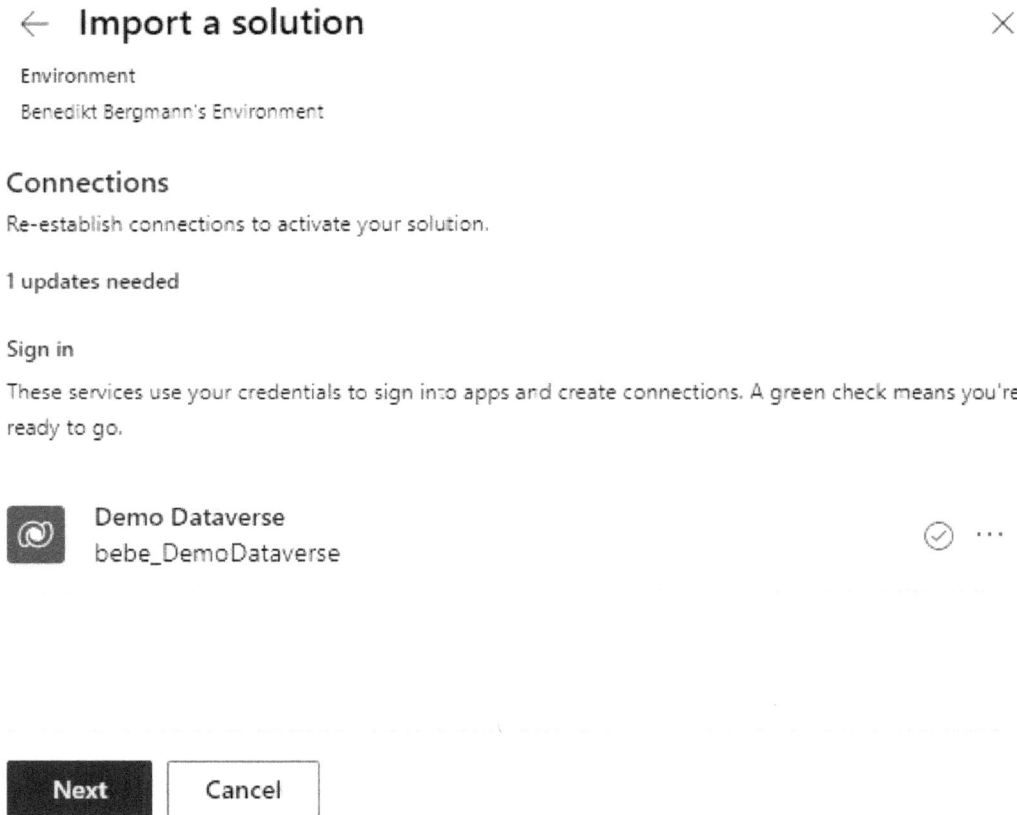

← **Import a solution**   ✕

Environment
Benedikt Bergmann's Environment

Connections
Re-establish connections to activate your solution.

1 updates needed

Sign in

These services use your credentials to sign into apps and create connections. A green check means you're ready to go.

Demo Dataverse
bebe_DemoDataverse   ⊘ ···

[**Next**] [Cancel]

Figure 4.31: Import Wizard Page 3 – Connection References

On the fourth page, we can set the correct values for the environment variables in the solution.

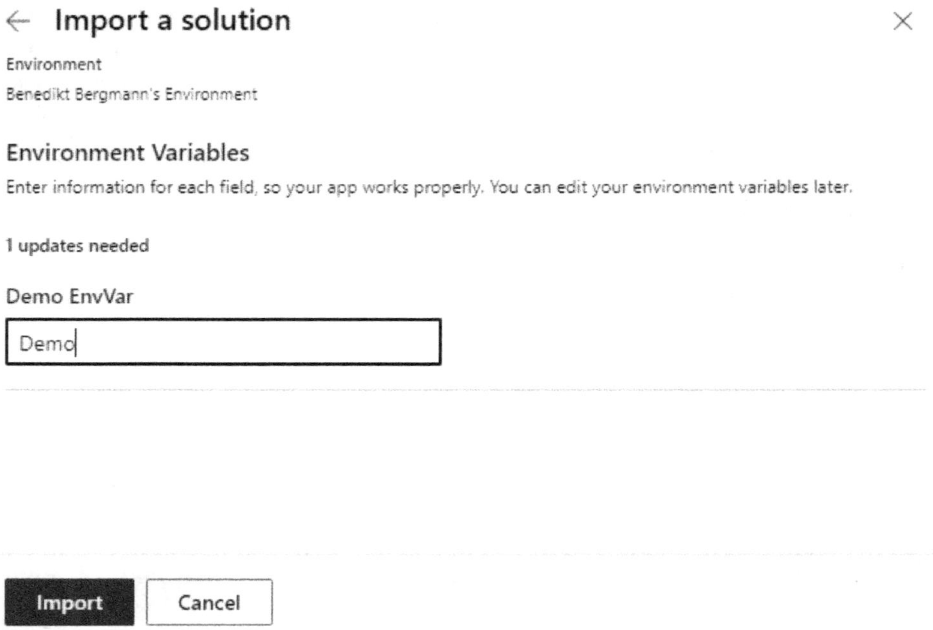

← **Import a solution**   ×

Environment
Benedikt Bergmann's Environment

Environment Variables

Enter information for each field, so your app works properly. You can edit your environment variables later.

1 updates needed

Demo EnvVar

| Demo |

Import   Cancel

Figure 4.32: Import wizard page 4 – Environment Variables

With a click on **Import**, the wizard will be closed and a notification will be shown that a solution import is ongoing.

> **One import at the time**
>
> Please note that there can only be one import active per environment at any time.

When the import is complete (either successful, with warnings or errors), the notification will change accordingly.

Summary

In this chapter, we have learned a lot about solutions. We have learned the basics, such as customization versus data, managed versus unmanaged solutions, update versus upgrade, and solution layering, but we also talked about solution segmentation and how to work with solutions, including how to handle components and how to export and import a solution.

This will help you handle any ALM process you might have, no matter whether it is a manual process or an automated process.

In the next chapter, we will learn about the Power Apps CLI.

Questions

1. Which kind of solution should we use in downstream environments?

   A. Managed

   B. Unmanaged

   C. Segmented

2. Which two options do we have to export a solution manually?

   A. Solution context menu

   B. CLI

   C. Ribbon menu

3. Which option do we have when removing a managed component?

   A. Remove from Environment

   B. Remove from Solution

   C. Delete from Environment

Further reading

- Solution layering: https://learn.microsoft.com/en-us/power-platform/alm/solution-layers-alm

- Publisher: https://learn.microscft.com/en-us/power-platform/alm/solution-concepts-alm#solution-publisher

5
Power Platform CLI

A **command-line interface** (**CLI**) is a program that accepts text input and executes commands accordingly. A lot of different CLIs do exist. Microsoft also has created a CLI for Power Platform.

This chapter will give you an overview of the Power Platform CLI and how to use it. We will cover the different command groups available as well as deep dive into the most important ones for a successful **application lifecycle management** (**ALM**) process in Power Platform.

In this chapter, we will cover the following main topics:

- Command Groups
- Solution Command Groups
- Installing the Power Platform CLI

Technical requirements

You can download the Power Platform CLI from the Microsoft website: `https://learn.microsoft.com/en-us/power-platform/developer/cli/introduction#install-microsoft-power-platform-cli`.

About the Power Platform CLI

In the beginning, CLIs were only used to execute operating system functions. Over the years, more and more CLIs for public APIs were created. The Power Platform CLI is one of those since it targets the APIs of Power Platform. It executes operations not against your local machine, but Power Platform.

The intention of the Power Platform CLI is to have a place where developers easily can execute different tasks in Power Platform.

> **Tip**
>
> The Power Platform CLI was previously called **Power Apps CLI**. This is the reason why the shortcut to run it in your command line is `pac`. Within the community, the Power Platform CLI is also referred to as the **PAC**.
>
> The PAC has a wide range of operations, spanning a large portion of governance and ALM areas within Power Platform.

Understanding the Power Platform CLI is a crucial part of mastering ALM for Power Platform. This is because, on one hand, the PAC is increasing the productivity of developers, and on the other hand, Power Platform Build Tools in Azure DevOps as well as GitHub Actions for Microsoft Power Platform do build on top of the Power Platform CLI.

Command Groups

Let's dive a bit into the different Command Groups currently available within the Power Platform CLI:

- `pac admin`: This handles the environment life cycle of your tenant.
- `pac auth`: This handles your connections to various environments and services.
- `pac canvas`: This operates with canvas apps and the associated `.msapp` files.
- `pac catalog`: This works with the new catalog feature.
- `pac connection`: This handles connections to Dataverse within your environment.
- `pac connector`: This manages custom connectors.
- `pac copilot`: This contains tools and utilities for use with Copilot.
- `pac data`: It contains commands for importing and exporting data from a Dataverse environment.
- `pac modelbuilder`: This generates an early-bound definition for the Dataverse API and tables.
- `pac org`: This helps in handling all the different organizations the connected user has access to.
- `pac packages`: This contains commands for working with Dataverse packages.
- `pac pcf`: The first command that was included in the PAC. It contains commands to work with the Power Apps component framework.
- `pac pipeline`: This is for handling pipelines in Power Platform.
- `pac plugin`: This contains commands for working with plugin class libraries for Dataverse.
- `pac powerpages`: This contains commands for working with Power Pages.

- `pac solution`: Commands to work with solutions. This is the area we will talk more about later in this chapter.

- `pac tool`: This is for installing Power Platform tools, such as **Plugin Registration Tool (PRT)**, the **Configuration Migration tool (CMT)**, and **Package Deployer (PD)**.

The next section will go into detail about the solution command group. This is important to understand since we will use those commands in creating the actual pipeline.

Solution Command Group

The commands within the Command Group solution are of special importance when it comes to ALM. This is mainly because, as we have learned in the previous chapter, solutions are the pillar of the ALM process within Power Platform. Therefore, it is important to understand how they can be handled.

We will take a deeper look at the following commands within this group:

- `export`

- `import`

- `unpack`

- `pack`

- `upgrade`

- `create-settings`

export

As the name suggests, this command is used to export a Dataverse solution from a target environment.

This command has one required parameter: `--name/-n`. We'll cover an example of this command at the end of this section.

--name/-n

Let's consider the name or schema name of the solution in question. This can be found in the second column of the solution list in the **Maker** portal. This can be accessed by selecting **Solutions** from the left-hand menu.

Display name ∨		Name ∨
ALM Solution	⋮	ALMSolution

Figure 5.1 - A Screenshot of the solution List

It also has some optional parameters, explained as follows.

-- async/-a

If present, the solution will be exported asynchronously.

--include/-i

A comma-separated list of the settings that should be included in `export`. Possible values are as follows:

- `autonumbering`
- `calendar`
- `customization`
- `emailtracking`
- `externalapplications`
- `general`
- `isvconfig`
- `marketing`
- `outlooksynchronization`
- `relationshiproles`
- `sales`

--managed/-m

When present, the solution will be exported as `managed`.

--max-async-wait-time/-wt

The maximum minutes the asynchronous job will run. The default is 60.

--overwrite/-ow

When present, the exported `.zip` will overwrite the file in the local system when it already exists.

--path -p

The output path of the generated solution `.zip` file.

An export command example

An example of an `export` command could be the following:

```
pac solution export --path c:\ALMSolution.zip --name ALMSolution
--managed true -async -overwrite
```

The shown command will export a solution called ALMSolution as managed in an async process. It will place the exported solution .zip file in the C: folder under ALMSolution.zip and overwrite any already existing file with the same name.

import

This command is used to import a previously exported solution to a downstream environment.

It only has optional parameters.

--activate-plugins/-ap

When present, import will attempt to activate plugin steps as well as workflows.

--async/-a

When present, import will be asynchronous.

--convert-to-managed/-cm

When present, import changes the components that are present in the solution to managed when they are present in the environment as unmanaged. It is only available when importing a managed solution. This parameter can be used to clean up an environment and is crucial when switching from unmanaged to managed in downstream environments.

--force-overwrite/-f

When present, all unmanaged customizations are overwritten in the target environment. It is only available when importing a managed solution. This parameter can be used to clean up an environment.

--import-as-holding/-h

When present, the solution will be imported as a holding solution. This will result in a second solution with the _Upgrade prefix in the system. This upgrade (or holding) solution has to be applied manually or through a separate step.

--max-async-wait-time/-wt

The maximum minutes the asynchronous job will run. The default is 60.

--path/-p

The path to the solution .zip file. If it is not specified, import assumes that the current folder is a cdsproj project.

--publish-changes/-pc

When present, a publish all customization is executed after the solution `import` succeeded.

--settings-file

The path to the `settings` file, which includes the values for connection references and environment variables. Later in this chapter, this will be explained a bit more.

--skip-dependency-check/-s

When present, the dependency check against the dependencies flag is skipped.

--skip-lower-version/-slv

When present, the solution `import` is skipped in case the same or higher solution is already installed in the target environment.

--stage-and-upgrade/-up

When present, `import` will do a combined stage and upgrade in one step.

An import command example

The following is an example of an `import` command:

```
pac solution import --path c:\ALMSolution.zip -async
```

Microsoft has improved the update import a lot in the last months. Importing a managed solution in update mode without `-convert-to-managed` and `-force-overwrite` can increase import speed tremendously.

unpack and pack

The `unpack` and `pack` commands are very similar to each other. The `unpack` command is used to extract solution components from a solution `.zip` file into an output folder on the local file system, whereas the `pack` command does the opposite. Both have the same parameters.

They have one required parameter: `--zipfile/-z`.

--zipfile/-z

The full path to the solution `.zip` file, which should be unpacked/packed, as well as the following optional parameters.

--allowDelete/-ad

When present, `unpack/pack` can delete files from the local file system.

--allowWrite/-aw

When present, unpack/pack can write files to the local file system.

--clobber/-c

When present, unpack/pack can delete or overwrite read-only files.

--disablePluginRemap/-dpm

When present, the plugin fully qualified type name remapping is disabled. We will talk about the remapping more in detail later in this book.

--errorlevel/-e

This defines the minimum level for log output. It can be one of the following values:

- Verbose
- Info
- Warning
- Error
- Off

The default is Info.

--folder/-f

The path to the folder where the solution .zip should be unpacked to/packed from.

--localize/-loc

When present, unpack/pack will extract all string resources into .resx files.

--log/-l

The path to the log file.

--map/-m

The path to the mapping XML file. unpack will not handle files present in the mapping XML file. The pack will take the files from the mapping folder instead.

--packagetype/-p

This defines the unpack/pack type. It could be one of the following values:

- Both
- Unmanaged
- Managed

The default is *Unmanaged*.

--processCanvasApps/-pca

When present, unpack/pack will also handle canvas apps.

--singleComponent/-sc

This defines whether unpack/pack should be performed only on a single component type. The following are possible values:

- WebResource
- Plugin
- Workflow
- None

The default is *None*.

--sourceLoc/-src

When present, the string resources of the presented language will be exported to a .resx file. You can use one of the following values:

- Auto
- **Language code identifier (LCID)**
- **International Organization for Standardization (ISO)** language code

--useLcid/-lcid

When present, unpack/pack will use LCID values instead of ISO language code for the generated .resx files.

--useUnmanagedFileForMissingManaged/-same

When present, unpack/pack will use the unmanaged XML file when there is no managed one.

An unpack command example

The following is an example of the unpack command:

```
pac solution unpack --zipfile C:\ALMSolution.zip --folder .\
ALMSolutionUnpacked\. --async
```

The shown example will unpack a solution .zip called ALMSolution.zip, which is stored in the C: folder. The process of unpacking will be executed as an async operation and store the extracted components in a folder called ALMSolutionUnpacked in the current working directory.

A pack command example

The following is an example of the pack command:

```
pac solution pack --zipfile C:\ALMSolution.zip --folder .\
ALMSolutionUnpacked\. --async
```

This pack command will pack the content of the ALMSolutionUnpack folder in the current working directory to a solution file called ALMSolution.zip. The pack command will be processed asynchronously.

upgrade

The upgrade command will apply a previously holding the installed solution.

It has one required parameter: --solution-name/-sn.

--solution-name/-sn

The schema name of the solution to upgrade, as well as two optional parameters.

--async/-a

When present, import will be asynchronously.

--max-async-wait-time/-wt

The maximum minutes the asynchronous job will run. The default is 60.

upgrade command example

The following is an example of the upgrade command:

```
pac solution upgrade --solution-name ALMSolution --async
```

The example will apply the holding solution to ALMSolution in an asynchronous process.

create-settings

This command will create a `settings` file from a solution `.zip` or `solution` folder.

It has three optional parameters.

--settings-file/-s

The path to the `.json` `settings` file to use.

--solution-folder/-f

The path to the `solution` folder to generate the file from, either to the root where the `Solution.xml` or `.cdsproj` file is stored.

--solution-zip/-z

The path to the solution `.zip` file to generate the file from.

create-settings command example

The following is an example of the `create-settings` command:

```
pac solution create-settings --solution-zip C:\ALMSolution.zip
--settings-file .\ALMSolution-Test.json
```

This example creates a `settings` file called `ALMSolution-Test.json` from our `ALMSolution.zip` file in the current folder.

Installing the Power Platform CLI

The Power Platform CLI can be installed in two different ways:

- Using Power Platform tools for Visual Studio Code
- Using Power Platform CLI for Windows

The difference between them will be accessibility. When we use the Visual Studio Code tooling, the CLI will only be usable in Visual Studio Code, whereas the version for Windows will make it usable even in any CMD or PowerShell window.

Power Platform tools for Visual Studio Code

To install the Power Platform tools for Visual Studio Code, follow these steps:

1. Open Visual Studio Code.
2. Go to **Extensions**.

3. Search for Power Platform Tools.

4. Select **Install**.

This will install the Power Platform tools for Visual Studio Code on your machine. Those include the PAC as well.

> **Important note**
> This is possible cross-platform for Windows, Linux, and macOS, and works for all versions supported for Visual Studio Code.

Power Platform CLI for Windows

For the standalone CLI, there are two different ways of installing it – either through dotnet tool or a .msi file:

-   To use dotnet tool, you need to have .NET installed. Then, execute the following command:

    ```
    dotnet tool install --global Microsoft.PowerApps.CLI.Tool
    ```

-   To use the .msi file, you have to download it from the Microsoft Power Platform CLI website and run the file.

> **Important note**
> This only works for Windows. Some Command Groups of the Power Platform CLI are only available through this version.

Summary

In this chapter, you learned what a CLI is as well as that there is a CLI for Power Platform.

You now understand which parts the Power Platform CLI (PAC) contains and the details of the most important ones when it comes to ALM:

-   Solution import

-   Solution export

-   Solution unpack/pack

-   Solution upgrade

-   Creating a settings file

This chapter also described how you can install the Power Platform CLI.

The next chapter will cover Environment Variables, Connection References, and data.

Further reading

For more information on the topic this chapter covered, please refer to the Microsoft documentation around the Power Platform CLI: `https://learn.microsoft.com/en-us/power-platform/developer/cli/introduction`.

Environment Variables, Connection References, and Data

This chapter is all about **Environment Variables (EnvVars)**, **Connection References (ConRefs)**, and configuration and reference data. These are crucial components of an ALM process. They make it possible to change the behavior of an implementation depending on which environment we are in. In addition, their values can be changed automatically depending on the environment we are deploying to. All of this together makes them important parts of a healthy ALM process and it is necessary to understand how they work.

This chapter will cover the following topics:

- Environment variables
- Connection references
- Configuration and reference data

Environment variables

First, we want to take a look at EnvVars. In this section, we will learn what EnvVars are, the different types we have at hand, how to create them, as well as how to use them.

EnvVars are a way of using different values depending on the environment the application is running in without the need to change the implementation in every environment or add if statements. This could, for example, be the email address of a recipient or the ID to a queue that is needed in a cloud flow.

Dataverse contains two tables for implementing EnvVars. One is **EnvVar Definition** and the other is **EnvVar Value**. Every definition has one value.

The following figure illustrates the table structures.

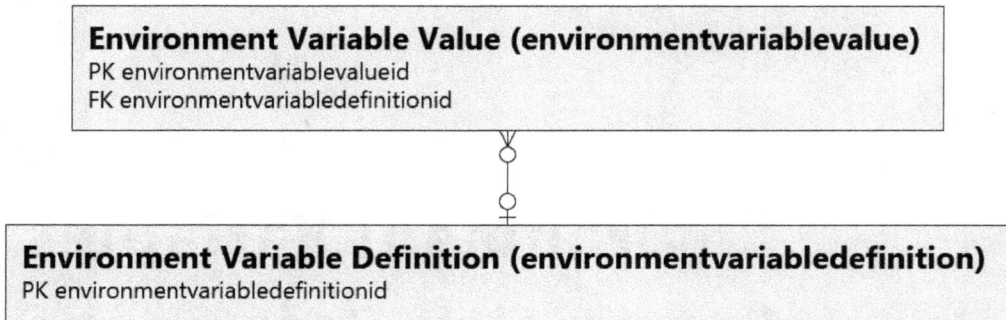

Figure 6.1: EnvVars table structure

The definition contains everything to specify what an EnvVar looks like. This is the schema name, display name, description, data type, and default value.

The value represents the current value of the EnvVar in a certain environment. It should not be included in a solution and should be set in a target environment either manually or through settings files. We will learn more about settings files in *Chapter 11*.

If a current value is present, it will be returned as the value of the EnvVar. Otherwise, the default value will be replaced.

Data types

When creating an EnvVar, we have to select which type it should be. Now we have six different types to choose from, as shown in this screenshot:

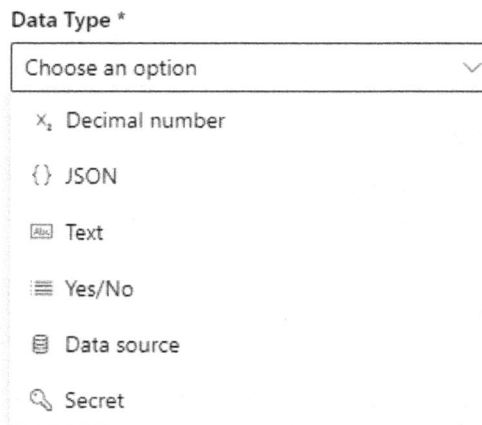

Figure 6.2: EnvVar data types

They are, namely, **Decimal number**, **JSON**, **Text**, **Yes/No**, **Data source**, and **Secret**, which we'll cover in the following sections.

Decimal number

As the name suggests, the variable contains a number with (or without) decimal places.

JSON

An EnvVar of this type could store a JSON object. This could be used to store more complex configurations that otherwise would require several EnvVars.

> **Max characters**
>
> It is important to mention that every EnvVar is limited to a maximum of 2,000 characters. That is especially important for EnvVars of the **JSON** type since the JSON object could easily exceed the character limit.

Text

This type of EnvVar would store a simple string. This is the most used type.

Yes/No

A **Yes/No** EnvVar represents a Boolean.

Data source

This type is a bit more complex. It represents a data source that can be used by cloud flows and canvas apps. If you create one EnvVar with this type, you have to select one of three connectors: either Microsoft Dataverse, SharePoint, or SAP ERP. It is different from connections, which we learn more about later in this chapter, in that it not only stores information on how to connect to a certain source but also describes what information it takes from the source. The SharePoint connector, for example, also stores the list, site, or document library that should be used.

> **Build Tools limitation**
>
> Power Platform Build Tools tasks are not able to handle this type of EnvVars in settings files yet. We will learn more about them in *Chapter 9* and about settings files in *Chapter 11*.

Secret

The last type is called **Secret**. It connects to Azure Key Vault and can be used to retrieve secrets or other secure information in a secure matter without the need to store it in plain text.

All users creating and using this EnvVar need access to the secret used in it.

> **Settings file connection string**
>
> When you want to set the value of an EnvVar of the **Secret** type, you have to present the complete connection string to the secret in question as a string. This is not as easy as setting the value through the UI.

Creating an environment variable

To create an EnvVar, we navigate to the maker portal and open a solution. After that, we select + **New | More | Environment variable**:

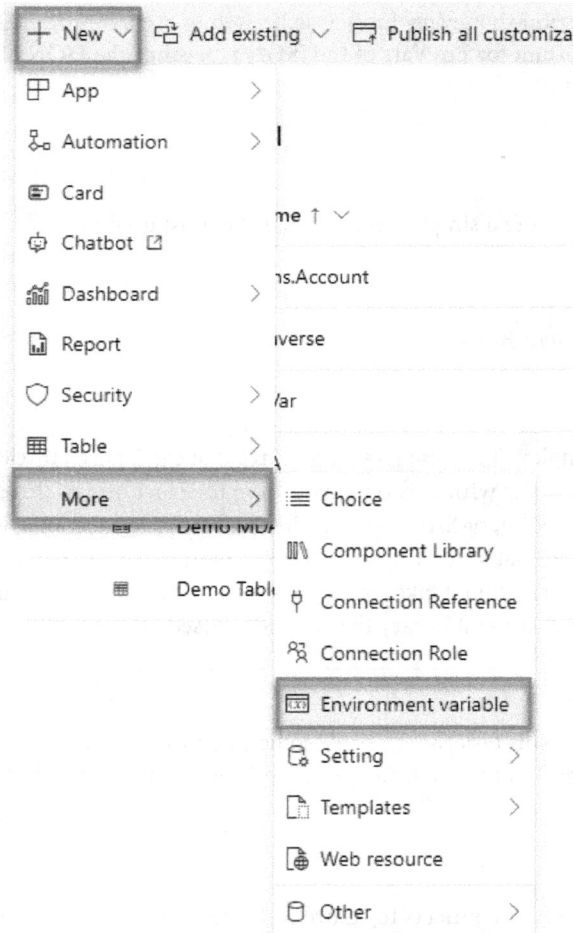

Figure 6.3: Create an EnvVar

This will open a pane on the right side. Here, we add all the values we would like to have. A recommendation is to directly add a current value by using the **+ New value** button underneath the **Current Value** section. This button will only be visible after we have selected the type.

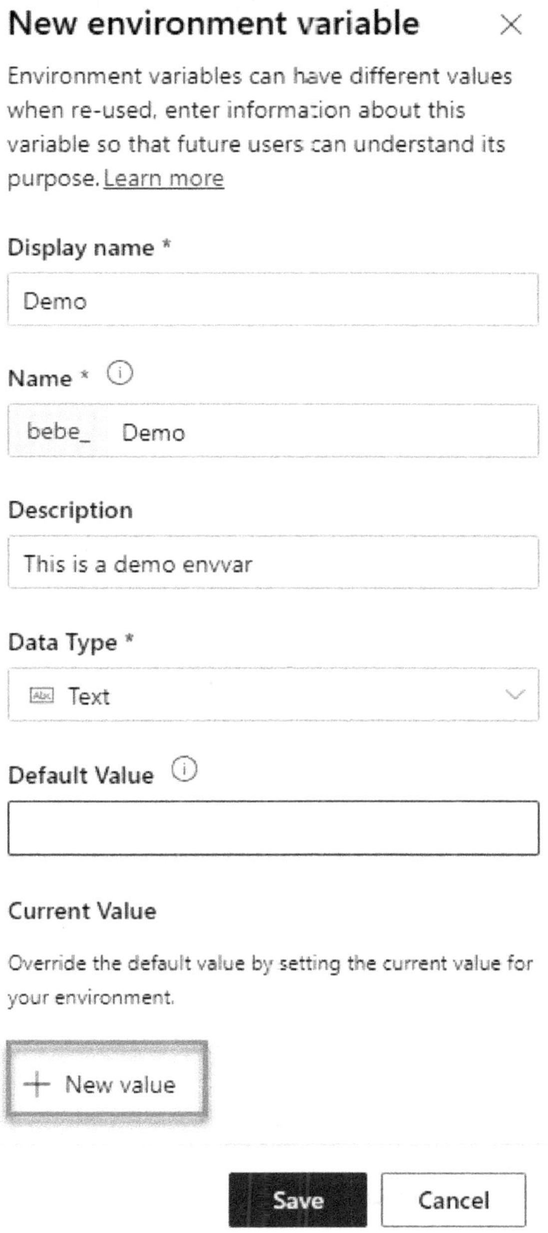

New environment variable ✕

Environment variables can have different values when re-used, enter information about this variable so that future users can understand its purpose. Learn more

Display name *

```
Demo
```

Name * ⓘ

```
bebe_    Demo
```

Description

```
This is a demo envvar
```

Data Type *

```
🔤  Text                                          ⌄
```

Default Value ⓘ

```

```

Current Value

Override the default value by setting the current value for your environment.

```
+ New value
```

Save   Cancel

Figure 6.4: Create an EnvVar and add a current value

The current value should not be included in solutions since it should only be used and valid for the current environment. As mentioned, the idea behind EnvVars is to be able to have different values for them in different environments. To make sure that the current value is not included when exporting a solution, we open the **Advanced** area and check that the **Export value** radio button is set to **No**.

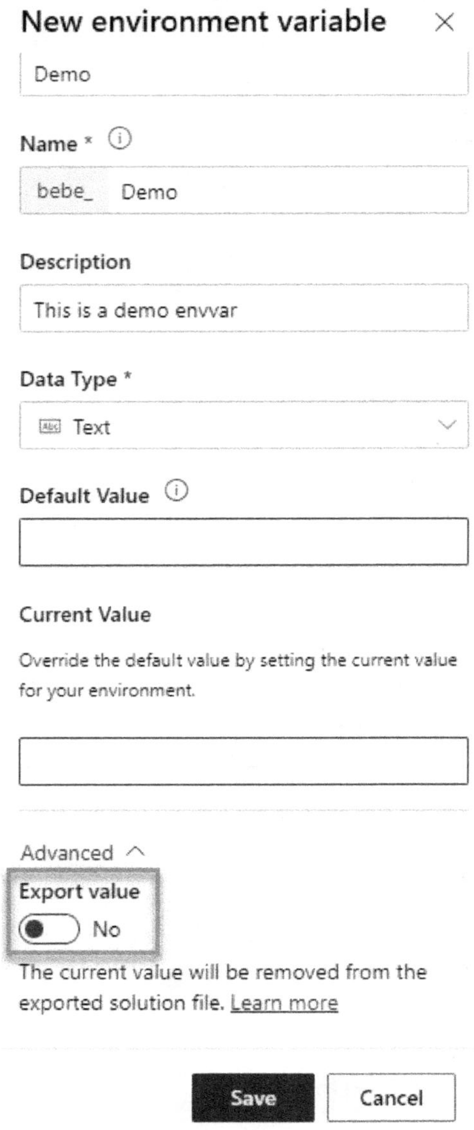

New environment variable ✕

Demo

Name * ⓘ

| bebe_ | Demo |

Description

This is a demo envvar

Data Type *

| Abc Text | ⌄ |

Default Value ⓘ

Current Value

Override the default value by setting the current value for your environment.

Advanced ∧

Export value

(●) No

The current value will be removed from the exported solution file. Learn more

Save   Cancel

Figure 6.5: Create an EnvVar – Advanced

With that, we have created our EnvVar, which we can now use in different parts of our implementation.

Applying an environment variable

An EnvVar can be used anywhere in our implementation. It could be used in cloud flows, plugins, custom APIs, or even Azure Functions.

Be aware that if you use EnvVars in custom code (for example, plugins, custom APIs, or Azure Functions), you would have to handle them manually.

In a cloud flow, the list of available EnvVars will be shown In the dynamic value popup. It is important to notice that the list will show all the EnvVars in the environment and will not be filtered on the solution you are working on. In some cases, the cloud flow UI will filter the list of EnvVars based on the type of the EnvVars and the required type of field you are trying to use the value in.

The following figure shows how this would look in the cloud flow UI.

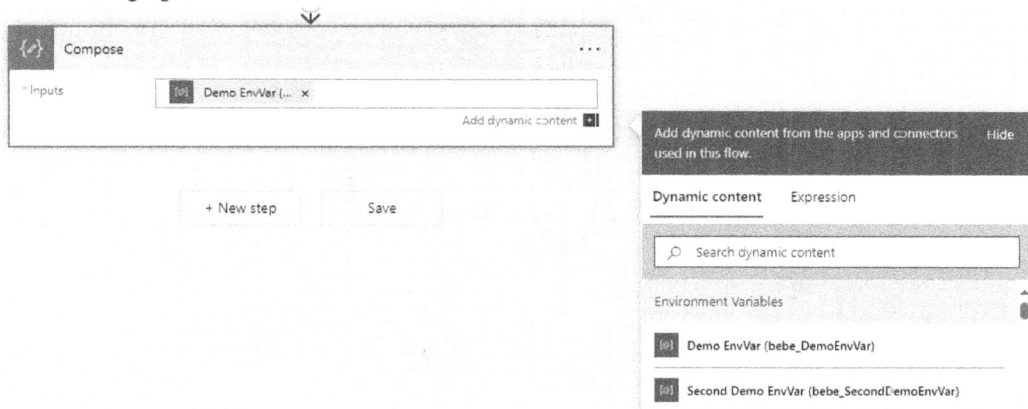

Figure 6.6: Using an EnvVar in a cloud flow

With that, we have learned what EnvVars are and how to create and use them. This is everything we need to know for now.

The next section will talk about connections and ConRefs. We will learn what they are and how to use them.

Connection references

ConRefs define which connector is used for a connection and are used in both solution-aware canvas apps as well as solution-aware flows.

Before the platform had ConRefs, a flow or canvas app would use connections directly. Since connections can't be included in a solution, they cannot be deployed with the component using it, which would break the component in downstream environments. The workaround was to go through all the steps in the flow or canvas apps where the connection was used and reconnect it. This would have to be done after every deployment.

By using ConRefs, cloud flows, and canvas apps don't need to use a connection directly. This makes it possible to change the underlying connection without the need to change the component, flow, or canvas app using it. The whole ALM process has become much easier and more manageable since ConRefs were introduced.

Let's look at its structure. Every step in a flow uses a ConRef. A ConRef defines which connector it is representing as well as which connection should be used. Each flow could have several ConRefs. Each ConRef uses one connector and one connection. Each connector and connection could be used in several ConRefs. The following diagram illustrates the structure.

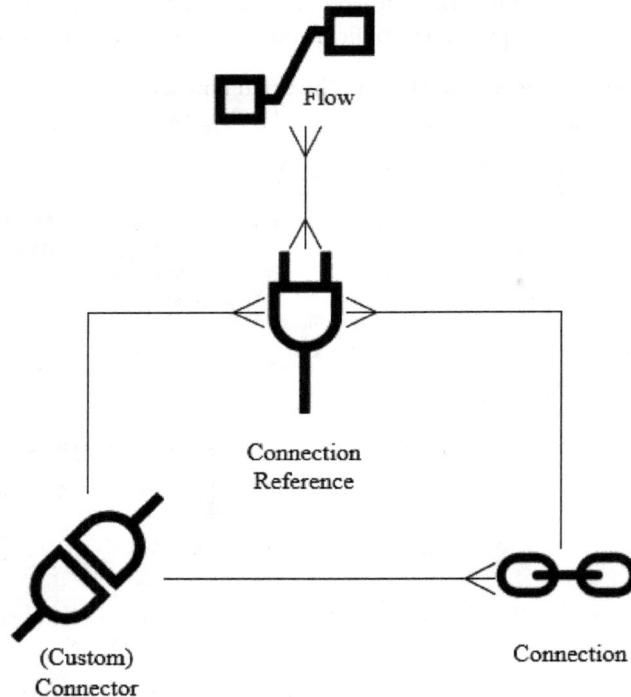

Figure 6.7: ConRef structure

> **Note**
>
> A recommendation is to use as few ConRefs to the same connector as possible. In the best-case scenario, only one ConRef is used per connector. There are exceptions where one would need more than one though. One example could be different integrations where we would like to know which surrounding system has made changes to the data in Dataverse. In that case, we would want to have two ConRefs so that we can use two different connections with different users.

The next subsection will explain connections and how they relate to ConRefs.

Connections

A connection represents everything needed to connect to a certain service. We can create connections to any of the 1,000+ Power Automate connectors or even custom connectors. They are used by Power Automate as well as other components in the Power Platform (such as canvas apps or low-code plugins).

A few of the connectors have the option to use a **Service Principal Name (SPN)** as a user to connect to the service in question. We will learn more about SPNs in *Chapter 9*. This has the following advantages over a named user:

- A service account would be the owner/modifier of the data

- The flow would be less coupled to a particular user

Every user has the option to create one or more connections to the same connector. This is especially useful when there is a requirement to know which integration has done which change. In that case, we could create several connections to the same connector using different SPNs.

Connections are not solution-aware and scoped to a user. Depending on the connector used, you might be able to share them with other users. We can always share them with SPNs. Make sure to share all connections with the SPN used in your pipelines. We will dive more into that later.

> Handling connections
>
> Since connections are scoped to users and we are not able to share all of them with users, a common workaround is to have one service account (basically a normal account that is shared between makers) to own the connections in production (and maybe some environment before that too). With that, everyone who needs access gets access to all connections (by logging in to the service account). This means we increase visibility into which connections are used where in our solution, as well as increasing maintainability since everyone can change the used connection if it is needed. In addition, we reduce dependency on user accounts since connections get deleted when an account is deactivated, which would happen when a person is leaving the company or project.

Tips and tricks

The last part of this section includes tips and tricks when it comes to connections and ConRefs.

Creating connections before deploying

It is highly recommended to create all the required connections before you start with your deployment. You are required to do so in an automated deployment process.

Creating connection references manually

It is better to create ConRefs manually and not through the cloud flow editor. This gives you the option to also decide the schema name and therefore apply a more consistent naming standard.

Renaming

A best practice is to give connections and ConRefs names that let a user know what they are used for. If they have been created automatically, you should rename them.

It is also recommended to use a consistent schema for naming them. I tend to use `<Area>` – `<Connector>` – `<Type>`. For a ConRef to Dataverse using a service principal that is used in our Demo app, it would, for example, be `Demo - Dataverse - Service Principal`.

Cleaning up connection references

At least every now and then, you should take the time to clean up all ConRefs. It is best to do it before every release. By clean up, I mean consolidating as many as possible, naming them correctly, and making sure the correct ones are used in the right places.

Service principal connections

A good practice is to use service principal connections whenever it is possible. The number of connectors able to use SPNs is slowly getting bigger. One of the most important connectors, Dataverse, already supports it.

Sharing connections

All connections used in downstream environments have to be shared with the service principal used in your pipelines. This is needed so that the import job can activate your flows.

Configuration and reference data

The last part of this chapter is about configuration and reference data. A solution does not contain actual data or rows in a table. There are several use cases where we would like to transport data during a deployment, though. For example, the configuration of a Power Pages instance is only done in non-solution-aware tables.

Another example is reference data for integration. With that, we mean data that represents certain information in the other system that we need to map into our system. This could be user-friendly names of drop-down options that are only delivered as integers in the integration or mappings between an input value from the integration and the corresponding record in our system. Those are usually the same in every environment, which means we would like to transport them during our deployment so that changes are reflected in the downstream environment as well.

There are different ways of transporting data in Power Platform, for example, using the **Data Migration Utility (DMU)** from Microsoft, the Data Transporter plugin from XrmToolBox, or Shuffle Runner.

My recommendation is to use the DMU since it is supported by Power Platform Build Tools and deployment packages. In addition, it was created by Microsoft and will stay around for another while at least.

To use it, we would connect the DMU to our development environment, create a schema file containing all the tables we would need, export the data according to the schema file, and import it to the downstream environment.

The following figure illustrates this process:

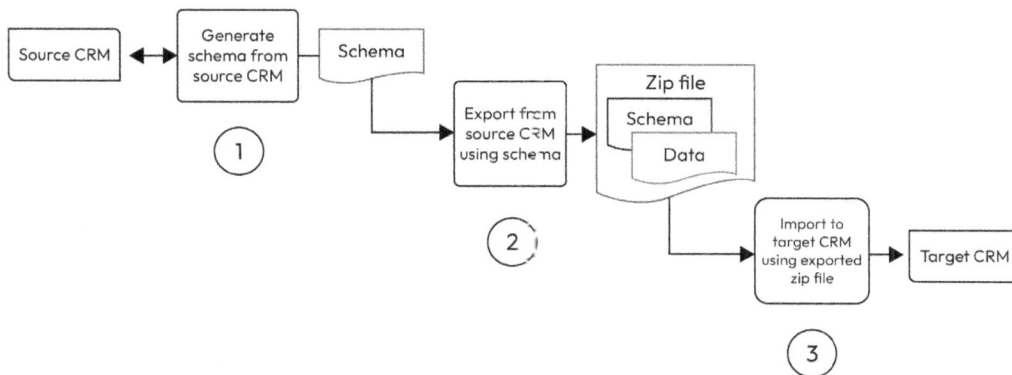

Figure 6.8: Flow for DMU

> **Better management**
>
> Power DevOps Tools created by Wael Hamze has a task that can split the data ZIP file into multiple files. This makes the management of the data in version control easier.

In *Chapter 11*, we will learn in detail how to create a schema file and how to use it in our pipelines to automatically export and import data.

Summary

In this chapter, we have learned the basics about three important areas when it comes to ALM: EnvVars, ConRefs, and transporting data.

We also learned how those areas work and are tied together, as well as how they can in theory be used in an ALM process. This is important base knowledge that we will expand on in the following chapters.

In the next chapter, we will deep dive into what the source code-centric approach is and its difference from the environment-centric approach.

Questions

1. How many environment variable types are there?

    A. 5

    B. 6

    C. 7

2. What is the preferred way of moving data in an automated manner?

    A. Shuffle

    B. Data Transporter

    C. Data Migration Utility

3. Which statement is correct about connections?

    A. They are solution-aware

    B. They are user scoped

    C. They can always be shared with other users

Further reading

- Data Migration Utility: `https://learn.microsoft.com/en-us/power-platform/admin/manage-configuration-data?WT.mc_id=DX-MVP-5002475`

- Shuffle Runner: `https://github.com/Innofactor/Innofactor.Crm.CI`

- XrmToolBox: `https://www.xrmtoolbox.com/`

- Data Transporter: `https://www.xrmtoolbox.com/plugins/Colso.Xrm.DataTransporter/`

- Power Automate connectors: `https://powerautomate.microsoft.com/en-us/connectors/`

7

Approaches to Managing Changes in Power Platform ALM

As is the case for all software development, for Power Platform projects, it is important to have a consistent way of managing your changes.

Besides a "normal" strategy for all code parts in a Power Platform, there are mainly two different approaches for configuration made in a Power Platform Development environment:

- Environment-centric
- Source code-centric

In this chapter, we will learn the difference between those two, as well as the environmental considerations when it comes to those mentioned approaches.

We will also look into branching strategies when it comes to Power Platform.

We'll cover the following topics in this chapter:

- Environment-centric approach
- Source code-centric approach
- Combining both of the approaches
- Branching
- Process of a source code-centric approach

In this chapter, we will talk about version control (or source control). It is an approach that is used to track and manage changes to software code. Source control is widely used in all software development, including Power Platform projects.

Environment-centric approach

The first approach we would like to introduce is the environment-centric approach, whereby the single source of truth is your development environment.

The basic process is that you export your solution as managed from DEV and deploy it to the downstream environments. This is the approach that is used in most of the current Power Platform projects.

Usually, this means that the focus is on the development environment. It will be nurtured and taken care of. This, in itself, has a positive and negative impact. When the development environment fails and becomes unusable or any wrong change is made that can't be undone, we might run into problems, since it is the only environment where we can make changes.

The export and deployment could be done manually or automated with pipelines.

The following figure illustrates the approach.

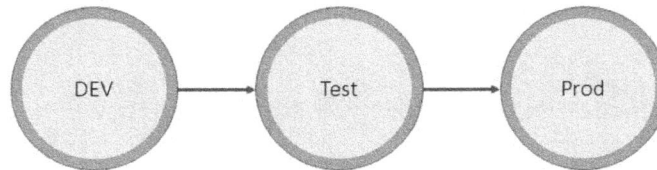

Figure 7.1 – Environment structure and flow for the environment-centric approach

Advantages and disadvantages

The environment-centric approach is easy to set up and manage. This also explains why it is, as mentioned, the one used in most of the Power Platform projects. In addition, it is the most natural one since, especially at the beginning of a project, there are a lot of changes made by a lot of (citizen) developers. At the same time, the time pressure is high, and usually, the focus is not on a proper **Application Life Cycle Management** (**ALM**) process. This leads to manual deployments, which basically require this approach.

Problems may arise when something happens to your development environment and it becomes unusable. In that case, you might lose the configuration you made so far, or at least need a considerable amount of time to restore your development environment.

As mentioned earlier, the development environment in this approach gets taken care of and nurtured a lot. This has another downside – it is getting "old". Since the development environment is so important, it basically never gets recreated. Therefore, every however-small configuration made to the environment stays as it is. Sometimes, these configurations are never revisited or documented. Also, the recreation is never "trained". As soon as there is some staff turnover information about at least some of this manual configuration is lost. All this together makes it even harder when the environment actually gets lost or corrupted somehow.

A good practice when it comes to software development in general is to recreate a development environment every now and then. For example, you could do this during every sprint or quarter. This is to avoid reaching a point where it is "old" or no one in the projects knows how to recreate it. This becomes, as described, hard with an environment-centric approach.

Source code-centric approach

The source code-centric approach is the second one we would like to introduce. As the name suggests, the source code (or repository) is your single source of truth in this approach.

An implementation would export the solution from development and unpack it to the repository. When the solution should be deployed to the downstream environment, the solution ZIP file is packed from the repository and deployed to the target environment.

This approach does require some kind of automation through pipelines. It is not really possible to do this manually. However, it is possible to introduce it after the environment-centric approach has been used earlier.

The following figure explains this approach. There is no direct relation between the environments. Everything moves through the repository.

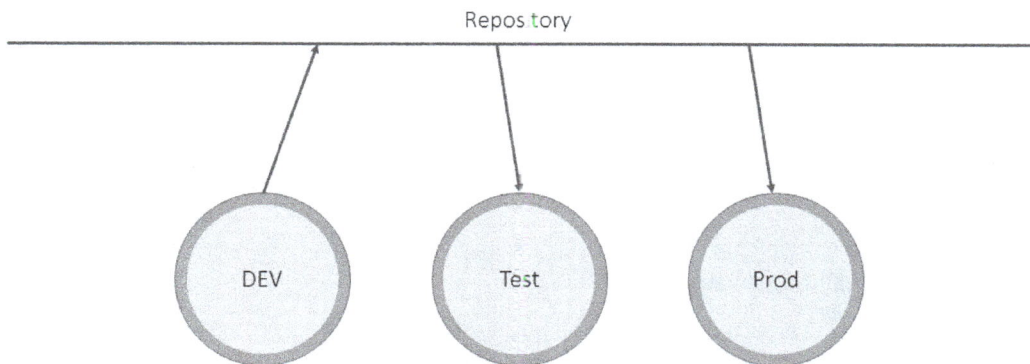

Figure 7.2 – Environment structure and flow of the source code-centric approach

Advantages and disadvantages

This approach does require more management and discipline. It also needs to be implemented in an automated approach via pipelines. Furthermore, the pipelines get more complex.

The dependency on certain environments is minimized. This means that if the development environment becomes unusable for any reason, it is possible to recreate it from the current version in the repository.

A branching approach, which is normally used when it comes to software development, can be implemented with this approach as well. This would make it easier for developers to start with Power Platform. We will go more into detail about this later in this chapter.

Combining both of the approaches

Often, project teams would like to use some parts and flexibility of the source code-centric approach without all of the added complexity. In most cases, this happens subconsciously, without the person even knowing there are the two mentioned approaches.

This can be achieved by a combination of both the source code-centric and environment-centric approaches. This means that the pipeline that does the export from your development environment will export a managed solution ZIP file but also unpack an unmanaged version of the solution to the repository. With that, we still have an environment-centric approach, but can easily recover from losing it since there is an unpacked version of the current solution in the repository. The mentioned problems around the "old" configuration would still be the same.

Branching

A plain development project usually utilizes some kind of **branching approach**.

For example, a lot of projects do use some variant of the Gitflow branching strategy. This contains the following branch types:

- **Main/Master**: As the name suggests, it is the main branch and the one that automatically gets created when the repository is created. Usually, it is either the working branch or the one related to Production.

- **Develop**: The Develop branch is used for day-to-day development. Usually, it contains the latest and greatest features that are ready to ship.

- **Feature**: A Feature branch is dedicated to a specific feature. Usually, it is used for bigger changes or implementations that take longer than a few hours and can't be completed in one workday. In that case, a developer works on their feature branch until the functionality is ready. That is when the Feature branch gets merged back into the Development branch. Often, there are several Feature branches open at the same time, usually by different developers.

- **Release**: Release branches represent the status of the code for a certain release. The code in that branch will be deployed to the downstream environments.

- **Hotfix**: Whenever there is a bug in production, a hotfix branch is created from the release branch that is currently installed in prod. The bug is fixed and then cherry-picked to the development branch.

Whenever a change is made in one branch (usually Feature or Hotfix), it is merged into the next "higher" branch using a **Pull Request (PR)**.

> **PR**
>
> A PR is a method of merging changes into a branch in a structured way. It is used to make sure code changes are quality-checked before they get applied.

These PRs should be protected with, for example, the following:

- An automated pipeline run that runs all automated tests (Unit, Integration, UI, etc.) can be used. Only when those run through successfully can the PR be merged into the other branch.

- At least one other team member should check the PR and make comments where necessary.

All this is used to improve the quality of the delivered software and is a common requirement in software development projects.

When it comes to Power Platform projects, this is nearly never required or done. The reason is mostly that it has not really been possible earlier. Since Microsoft has made a significant investment (and with that, improvement) in pro-dev tooling, as well as the underlying platform in general, this now is possible to achieve.

As briefly mentioned earlier, with the source code-centric approach, such branching and merging are possible. There could be different branches mapped to different Power Platform environments.

Let's say there is a feature request for a big restructuring of tables within the Power Platform solution we develop. Since the implementation of this feature will span over several sprints and we don't want to stop all other development, we could create a new Feature branch, which will be branched from the Development branch. This branch would be deployed to a separate Power Platform feature environment. Here, we could implement the necessary changes. When everything is ready, the Feature branch will be merged with our Development branch. With the next recreation of the Power Platform development environment, the changes would be in our common development environment.

Another scenario would be development environments for every developer or team.

Obviously, this requires a certain amount of management, as well as skills. It is most suitable for bigger projects. For smaller projects, the cost of the emerging overhead would not justify the advantages.

In addition, there are other things on the platform that don't completely work with this approach. For example, the unpack will generate so-called **noisy diffs**. This means that even if there was no change, Git would detect a change, since sometimes, the order of things in certain files (especially `Solution.xml`) isn't consistent.

Process of a source code-centric approach

Let's discuss what the process of a source code-centric approach would look like.

This approach needs three different pipelines:

- **Export pipeline**: The Export pipeline would, as the name suggests, do an export of the current state of the solution from your development environment. It should be built to allow it to run fast so that a user can run it whenever a sub-process is implemented. This could potentially happen several times a day.

- **Build pipeline**: The Build pipeline would take the unpacked solution from source control, pack it, and create an artifact that would be used for deployment to the downstream environments.

- **Release pipeline**: The Release pipeline takes the artifact of the build pipeline and deploys it to downstream environments.

In the following sections, we will look at which exact steps would be needed to implement the minimal version of the mentioned pipelines. *Chapter 11* of this book will give you more insights into how you can make those more complex and robust.

Export pipeline

As mentioned earlier, the Export pipeline should be as simple as possible so that it can run as fast as possible.

The following steps would be needed:

1. **Install tooling**: As with all pipelines running some steps for Dataverse, we have to make sure that the agent running the pipeline has the needed tooling installed. Both the Azure DevOps Power Platform Build Tools and the GitHub Power Platform Actions have steps for that.

> **Azure DevOps Agent Pool**
>
> This step could be skipped if you have your own Azure DevOps Agents, and if you have installed the latest version of the tooling yourself.

2. **Publish customizations**: Next, we have to publish all customizations to make sure that all changes will be reflected in the export we do.

3. **Export as unmanaged**: The third step is to export the solution as unmanaged.

4. **Export as managed**: After exporting the solution as unmanaged, we also have to export it as managed.

5. **Unpack**: When we have exported our solutions both as managed and unmanaged we can unpack both to our repository.

> **Canvas apps**
>
> With the current state of tooling for canvas apps, Microsoft does not recommend unpacking those if we are aiming to also pack the solution from Source Control. If you do, there might be errors with it along the way. There are workarounds for that; those would exceed the scope of this book.

6. **Commit**: Lastly, we have to commit our changes to the repository.

Build pipeline

The Build pipeline, in its simplest form. is a very small pipeline. When used in production, it should contain more complexity in the form of versioning and quality gates. We will talk more about those topics in *Chapter 11*.

The following steps are needed:

1. **Install tooling**: As already mentioned, with this step, we will make sure that the needed tooling is installed on the Agent executing the pipeline.

2. **Pack from repository**: The second step is to take the unpacked version we have stored in our repository and pack it into a solution ZIP file.

3. **Publish artifacts**: In the last step, we will publish the solution ZIP file as a pipeline artifact. Thus, the Release pipeline can deploy it to the downstream environment.

Release pipeline

The complexity of the Release pipeline very much depends on the setup and complexity of the project. For every environment the pipeline should deploy to, the following steps need to be executed:

1. **Install tooling**: As before, we have to install the tooling.

2. **Import solution**: The second step is to import the solution that we got as an artifact from the Build pipeline.

3. **Apply upgrade**: If the import step has installed the solution as a holding solution, this upgrade has to be applied.

Summary

There are two valid approaches to change management in Power Platform: environment-centric and source code-centric.

We explained the different approaches and discussed the advantages and disadvantages of both of them.

We also described that there often is a combination of both approaches used.

Lastly, we discussed branching when it comes to Power Platform.

In the rest of this book, we will mostly talk about the source code-centric approach and how to create automation around it.

The next chapter will describe the different tools and services we can use to automate the ALM process with Power Platform.

Questions

1. Which of the following is a benefit of the source code-centric approach?

   A. It is easy to manage

   B. Dependency to certain environments is minimized

   C. The single source of truth is your development environment

2. Which branch is often related to Production?

   A. Main

   B. Feature

   C. Development

   D. Release

3. What is the main benefit of an environment-centric approach?

   A. It is easy to manage

   B. Dependency to certain environments is minimized

   C. The single source of truth is your repository

Further Reading

Git Flow: https://www.gitkraken.com/learn/git/git-flow

8

Essential ALM Tooling for Power Platform

Up until this point, we have discussed the basics of **Application Lifecycle Management** (**ALM**) in general, as well as the specifics of the Power Platform. This chapter will explain the different tools we have to run our automated Power Platform deployments.

This includes an overview of pipelines in Power Platform, the ALM Accelerator, GitHub, and Azure DevOps. Azure DevOps and GitHub are tools that support more or less the whole of ALM, as discussed earlier. The ALM Accelerator and pipelines in Power Platform only support the automated deployment part of an ALM process.

We'll cover the following topics in this chapter:

- Pipelines in Power Platform
- ALM Accelerator
- Azure DevOps/GitHub
- Selecting the appropriate tool

Pipelines in Power Platform

Let's start with pipelines in Power Platform. The goal of pipelines in Power Platform is to democratize ALM in Power Platform and Dynamics 365. This is accomplished by giving ALM automation capabilities directly to platform makers. With that, it becomes more accessible and approachable for makers, administrators, and developers.

Before pipelines in Power Platform were introduced, adopting a healthy, automated ALM process for a project required a lot of domain knowledge and effort.

With pipelines in Power Platform, the following can happen:

- Admins can easily configure deployment pipelines

- Makers can use those pipelines in a UI experience they are used to

- Professional developers can extend those pipelines, as well as trigger them through the CLI

The capabilities provided by pipelines in Power Platform are similar to what we have with manual exporting and importing to move solutions to downstream environments. This is in case we don't extend these capabilities by using pre- and post-deployment customizations. With those, we can insert approval processes, or any other custom step we would like, into our pipeline.

Pipelines in Power Platform are deployed to every environment and can be used by anyone, which might lead to deployment efforts being made by makers without the governance team knowing about it. But again, this would be the same with manual exporting and importing.

After the initial setup and pipelines are configured (we will learn more about this in *Chapter 9* and *Chapter 10*), the deployment can be triggered from within a solution.

The following screenshot shows how this would look in the maker portal UI.

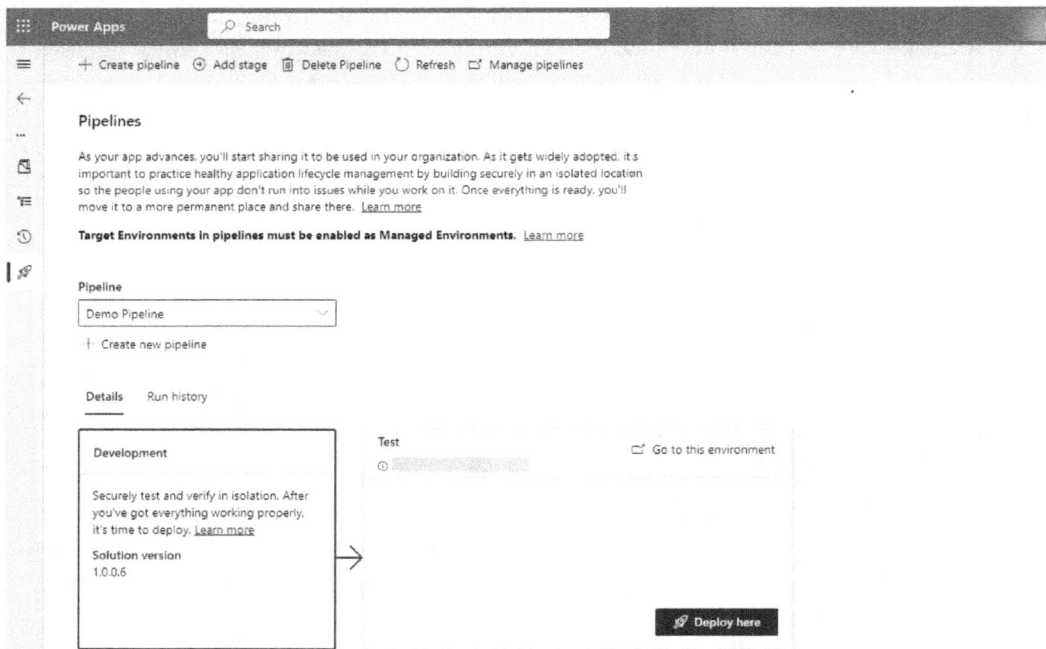

Figure 8.1: Pipelines in Power Platform

This also brings us to one of the limitations - pipelines in the Power Platform are executed on a solution level. This means if your project contains several solutions, you will need to know the order in which you have to deploy them, as well as wait for the next deployment until the current one is ready – once again, exactly as it would be with manual deployments.

Another downside is that pipelines in Power Platform, in their basic setup, only take Power Platform components into consideration. This means that you must handle the deployment of other components, such as Azure Functions, of your implementation separately. Thanks to the pre- and post-deployment customizations mentioned, this can be handled seamlessly but will require customization.

Organizations or projects that want to get started with ALM quickly and don't have a complex solution structure should choose pipelines in Power Platform.

ALM Accelerator

The second tool we will look at is the **ALM Accelerator** for Power Platform. It is a canvas app that is intended to provide a simplified interface on top of Azure DevOps Pipelines and Git source control in Azure DevOps. It is aimed at both low-code developers and code-first developers.

The ALM Accelerator relies on several different parts. It only works in combination with Azure DevOps, where it needs projects to store the template pipelines it ships, run the pipelines, and store solution artifacts and everything needed for a healthy ALM process.

It also contains the Dataverse solution, which needs to be installed in a host environment. The solution contains one **Model-Driven App (MDA)**, as well as a canvas app. The MDA is the admin app where one can configure everything that is needed. The canvas app is the one the developer uses to deploy their solutions.

To set up the ALM Accelerator, one needs a lot of permissions in Azure DevOps, Microsoft Entra ID, and the Power Platform. Most likely, you will need the help of a global admin to execute all the steps to install the app.

The following screenshot shows a zoomed-in version of the canvas app after the initial setup has been completed (we'll cover the setup part in *Chapter 9*).

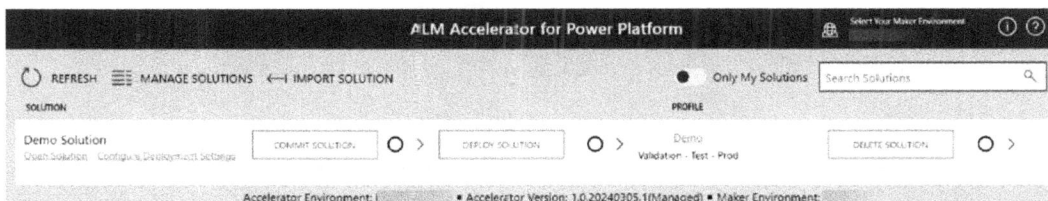

Figure 3.2: ALM Accelerator canvas app

The idea is that the ALM Accelerator presents a standardized way of deploying solutions from one environment to another. It follows best practices created by Microsoft and is intended to make life easier for admins. This is because they don't have to invent the wheel again and start from scratch to set up a harmonized ALM process. They can rely on the experience of Microsoft and reuse what has already been created and used successfully.

A downside is that the setup is not as flexible as creating everything from scratch to exactly match the project's needs, even though the ALM Accelerator can be extended, and to a certain degree changed, to fit the specific requirements.

Since the ALM Accelerator and all its components (including the used pipelines) are open source on GitHub, they can also be used as a reference to create your own implementation based on it.

The ALM Accelerator will only deploy Power Platform components. If your project includes surrounding components, such as Azure Functions, you will have to handle the deployment of those separately.

Usually, organizations who want to have control over their ALM process but don't have the resources to create everything from scratch choose the ALM Accelerator.

At some point in the future, the ALM Accelerator will not be developed further, since all the functionality has moved into pipelines in the Power Platform.

The ALM Accelerator is developed by the PowerCAT.

> **PowerCAT**
>
> **CAT** means **Customer Advisory Team**. Microsoft has different CATs for different areas. The one responsible for the Power Platform is called the PowerCAT.

Since the PowerCAT is not part of the product team directly, not everything they produce, such as the ALM Accelerator and the Creator Kit, is supported by Microsoft Support. If there is an issue, PowerCAT will offer support through the related GitHub repo.

Azure DevOps/GitHub

The last group of tools we would like to introduce are basically all ALM tools. As examples, since these are mostly used for Power Platform projects, we will talk about Azure DevOps and GitHub.

Both Azure DevOps and GitHub are tools outside of the Power Platform but are owned by Microsoft. Most projects using either of the tools are not Power Platform projects. Azure DevOps and GitHub have a lot more functionality than just automated deployment by running pipelines/workflows. That functionality includes the following:

- Requirement gathering
- Requirement prioritization
- Release planning
- Bug tracking
- Source code handling
- Testing
- Project documentation
- Different access levels
- Reporting

The following screenshot shows the overview page of an Azure DevOps project to illustrate the different functionalities available at a very high level.

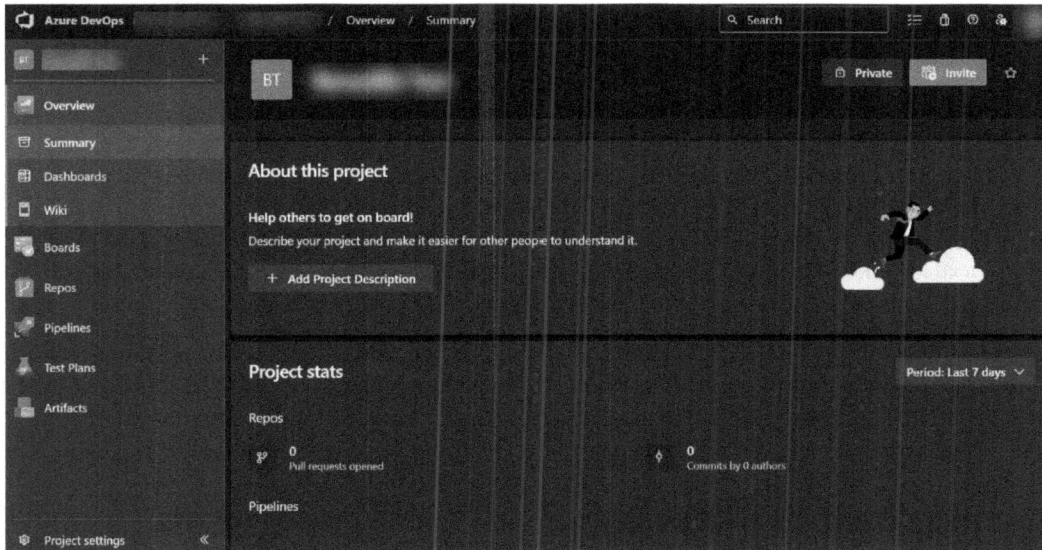

Figure 8.3: Azure DevOps project overview

The next screenshot shows a high-level overview of the functionality of a GitHub project.

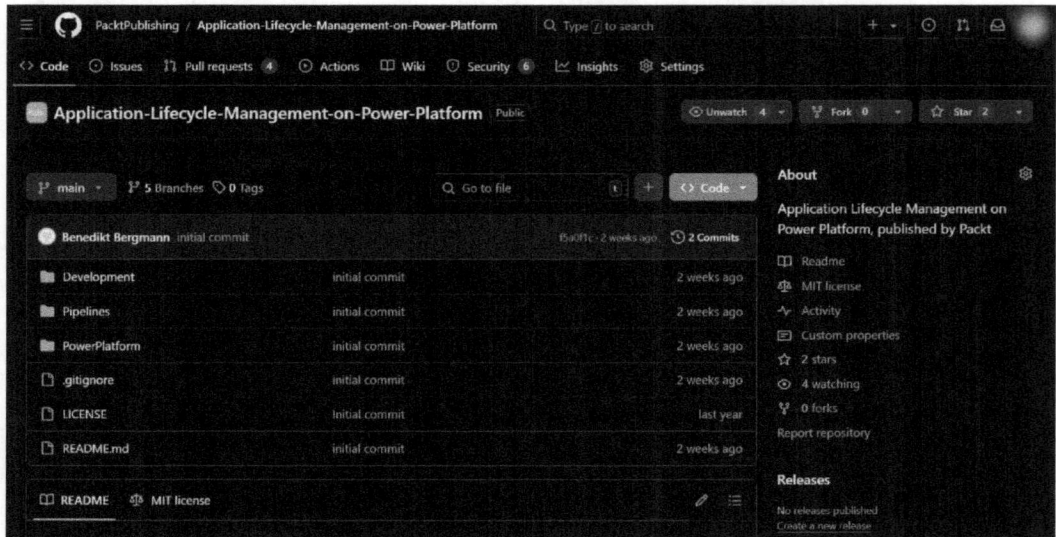

Figure: 8.4: GitHub project overview

For Azure DevOps, as well as GitHub, Microsoft has created first-party build tools/actions to interact with the platform. Those make it easier to implement a healthy ALM process.

> **Other ALM tools**
>
> You could implement a proper ALM process for Power Platform with basically any ALM tool. You are not restricted to any of the four (ALM Accelerator, Power Platform Pipelines, GitHub, or Azure DevOps) tools mentioned before. Through the Power Platform CLI (PAC), all the steps needed can be run in any place by creating custom Shell scripts using PAC commands.

An organization that already uses Azure DevOps, GitHub, or any other "full-blown" ALM tool would also use it for Power Platform projects. Another reason would be that you require granular control over what should happen in your deployment as well as having components in your implementation outside of Power Platform and resources at your fingertips to actually create the needed pipelines.

Selecting the appropriate tool

Now that we know about the different options, it is important to know when to select which of them. There is no tool that fits all projects or situations. It always depends on the requirements, knowledge level, the people within a project, and many other factors.

The following table answers the question about which tool to select. The table may not be exhaustive and may change over time, depending on the new capabilities of the different options.

Functionality	DevOps/GitHub	Pipelines	ALM Accelerator
Pro-dev required	Required	Not required	For setup
Support	Microsoft Support	Microsoft Support	Power CAT through GitHub issues
Quality control	Yes, configuration	Extensions	Extensions
Source code integration	Yes	Not at the moment	Yes
"In-product"	No	Yes	Partly, uses the Canvas app
Code-first development	Yes	No	Yes

Table 8.1: Comparison table

The preceding table, together with the information provided in the prior sections, should give you a good base to make an informed decision on which of the tools you should choose for your project.

Summary

We have learned about the different tools there are to implement a healthy application lifecycle management process for Power Platform.

We learned about the capabilities and functionality of the ALM Accelerator, Pipelines in Power Platform, GitHub, and Azure DevOps.

The ALM Accelerator and Pipelines in Power Platform (in their basic setup) are a good starting point for smaller projects where no surrounding components (such as Azure Functions) are included.

GitHub and Azure DevOps are also capable in other areas of the ALM process than just the deployment part, but they do require more investment when it comes to the setup and they have a different UI than makers are used to.

In the next chapter, we will see how to actually set up our project and all of the tools we have learned about in this chapter.

Questions

1. With which tool can you deploy Azure Functions?

    A. ALM Accelerator

    B. Azure DevOps

    C. Pipelines in Power Platform

2. Which tool requires all users in the environment to have premium licenses?

    A. ALM Accelerator

    B. Azure DevOps

    C. Pipelines in Power Platform

Further reading

- Pipelines in Power Platform: `https://learn.microsoft.com/en-us/power-platform/alm/pipelines`

- ALM Accelerator: `https://learn.microsoft.com/en-us/power-platform/guidance/alm-accelerator/overview`

- GitHub: `https://docs.github.com/en/get-started/start-your-journey/about-github-and-git`

- Azure DevOps: `https://azure.microsoft.ccm/en-us/products/devops`

9
Project Setup

Before we dive deeper into the actual implementation of automated **Application Life Cycle Management** (**ALM**) using pipelines, we have to shed some light on how one can set up the different systems needed.

In this chapter, we will learn how to set up **Azure DevOps** (**ADO**), as well as configure managed environments, the ALM Accelerator, and **Power Platform Pipelines** (**PPP**). In addition, we will talk about what the folder structure for a project could look like and which important areas it should cover.

We'll cover the following topics in this chapter:

- Power Platform Build Tools
- Setting up Azure DevOps
- Setting up GitHub
- Configuring managed environments
- Configuring the ALM Accelerator

Service principal

First of all, we have to create a **Service Principal Name** (**SPN**), or just service principal, which will be authenticated against our Dataverse environments.

In general, the recommendation is to use one SPN per Dataverse environment when it comes to integrations. This is so that our integration does not, by mistake, connect to the wrong environment. In the best case, a developer does not have access to the credentials for the SPN for the production environment.

> **Note**
>
> The Power Platform CLI also has a command to create a service principal for you. Read more at `https://learn.microsoft.com/en-us/power-platform/developer/cli/reference/admin#pac-admin-create-service-principal`. In addition, one could create SPNs through PowerShell as well.

When it comes to pipelines, we usually just create one SPN for a pipeline to lower the amount of SPN in Entra ID, which also makes it easier to maintain.

To create a service principal, we have to either log in to Azure Portal and navigate to **Entra ID**, or log in to Entra ID directly. There, we navigate to **App registrations** and create a new app registration.

> **Note**
>
> This action requires some privileged permissions. Depending on how your organization is set up, you might have to get help from an admin or request the new SPN via an internal order to the IT department.

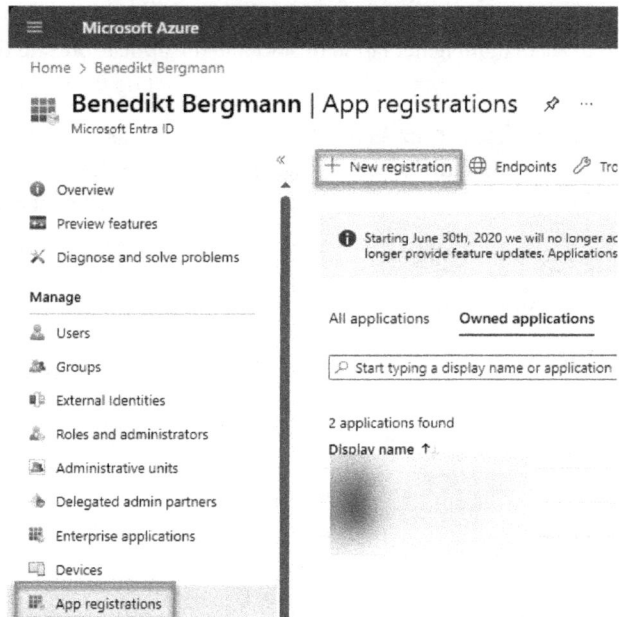

Figure 9.1 – Navigating to App registrations

After giving it a good and meaningful name, it is important to create a secret.

> **Tip**
>
> I usually use the `<customerPrefix>_<Area>_Pipeline` schema. There might be some naming conventions established (which may even be enforced) in your organization.

To create a named secret, navigate to the **Certificates & secrets** configuration within our newly created app registration and then click on + **New client secret**. Here, a new secret has to be created.

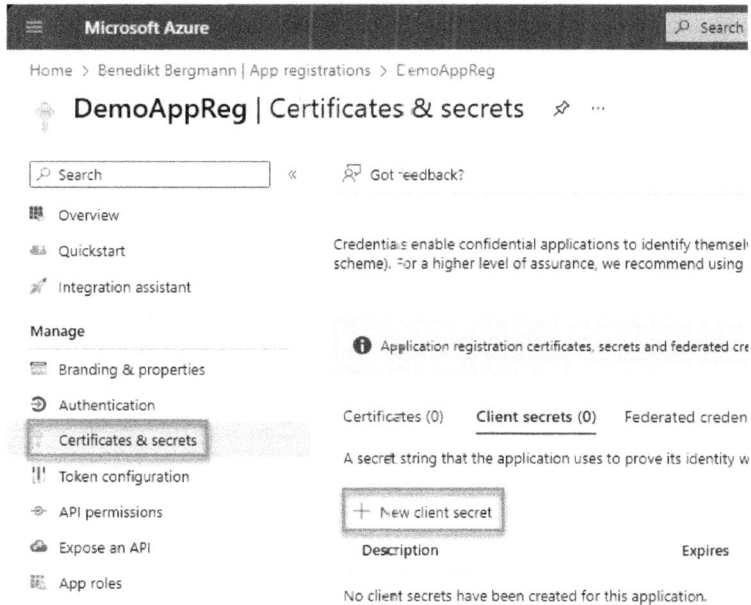

Figure 9.2 – Adding a secret to the app registration

> **Note**
> The secret value will only be shown once. It is crucial to copy this value since it is the one we will need later.

The Azure Portal only allows a maximum lifetime of 24 months for your secret. If you want to create secrets that last longer than this, you can achieve this through a CLI.

> **Tip**
> Read more about secrets that last longer than 24 months at `https://marketplace.visualstudio.com/items?itemName=WaelHamze.xrm-ci-framework-build-tasks`.

Now that we have created our new SPN, we have to authenticate it against Dataverse.

Add SPN to Dataverse

First, navigate to the **Power Platform Admin Center (PPAC)** (`https://admin.powerplatform.microsoft.com`).

After selecting the environment (usually DEV, Test/QA, and Production), go to the **Settings** page, and then to **Application users**. Here, we can add a new application user by clicking + **New app user**. In the side pane that comes up, first, select **App registration** from a list of app registrations and add an appropriate security role.

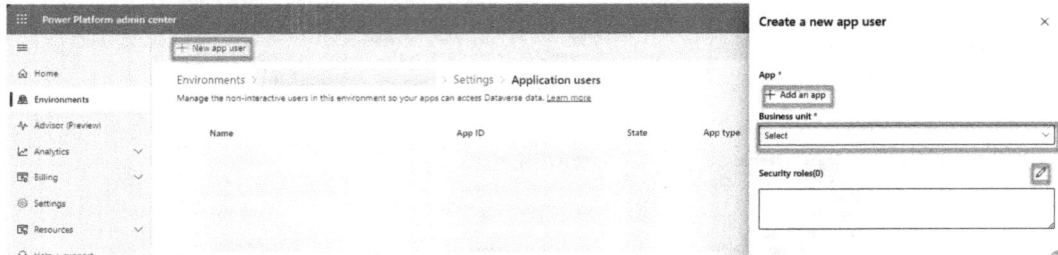

Figure 9.3 – Adding new app registration to Dataverse

> **Note**
>
> When it comes to security roles, a good recommendation is to create new security roles per integration to only give the level of access that's really needed. For pipelines, we should use the system administrator role though. This is because the pipeline has to have access to everything to be able to install all components while importing a solution.

This has to be done for every environment our pipeline is supposed to access.

The next section will describe what has to be set up in ADO to be able to run a pipeline successfully. This includes settings and permissions.

Setting up Azure DevOps

To get started, there are some things we have to do within ADO. Some of the following configurations are **ADO Organization**-wide and some have to be done per project or even repository.

> **Note**
>
> In some cases, you might want to use the same ADO project and still target different pipeline streams. For example, there could be one ADO project called **Power Platform Initiative (PPI)**, which should contain all development related to Power Platform. It might be part of a bigger organization-wide setup. Within this PPI project, we have to handle apps for different departments (HR, finance, marketing, and so on). All those departments would have separate Dataverse environments and repositories in the PPI ADO project.

Installing Power Platform Build Tools

The first thing we have to do is install the Microsoft Power Platform Build Tools.

> **Tip**
>
> Those are the official tools to use for pipelines connected to Dataverse and Power Platform. They have all the steps needed for a basic and healthy ALM process. When the process needs to be more complex (for any reason), there might be scenarios that aren't possible with the official build tools yet. In those cases, we either have to revert to custom PowerShell scripts or use the build tools provided by the community.
>
> The following is a link to the most popular build tools created by the community, Power DevOps Tools: `https://marketplace.visualstudio.com/items?itemName=WaelHamze.xrm-ci-framework-build-tasks`
>
> The official Microsoft Power Platform Build Tools can be found at `https://marketplace.visualstudio.com/items?itemName=microsoft-IsvExpTools.PowerPlatform-BuildTools`.

The installation is easy. Once you have found the Build Tools you would like to use in the ADO Marketplace, you have to hit the **Get it free** button (as seen in the following screenshot).

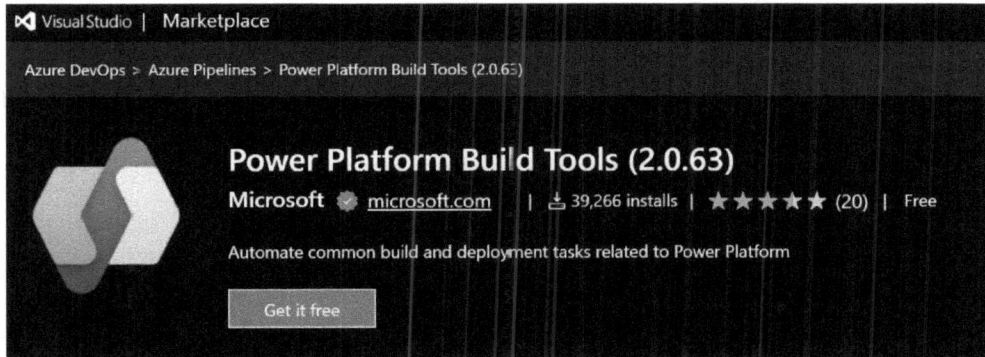

Figure 9.4 – Power Platform Build Tools in the Marketplace

This will redirect you to an installation screen for your organization's ADO. Since this step requires admin privileges in ADO, it will most likely show you a request screen.

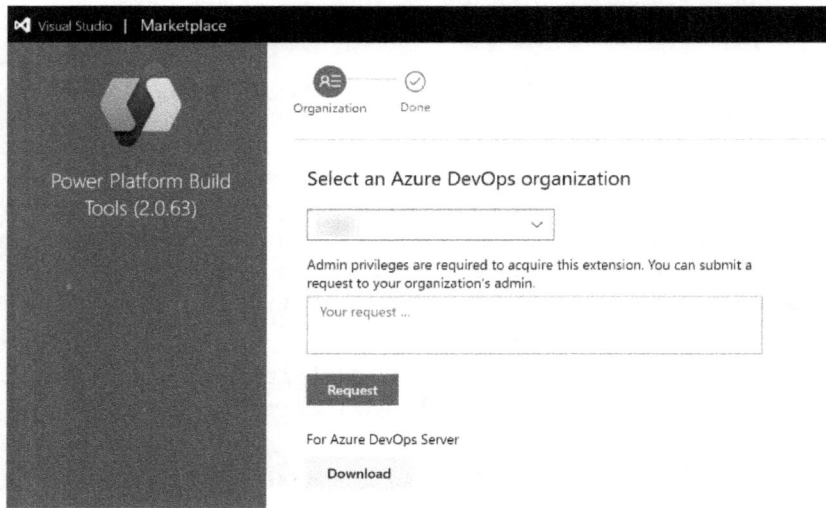

Figure 9.5 – Request screen to install Power Platform Build Tools

Here, we can select the correct organization and present some reasons to the admin for why we would need this to be installed.

On the same screen, we can download the current version of the Build Tools in case ADO is hosted on-premises.

In case you do have the correct permissions, you should see an **Install** button and can then proceed with the installation.

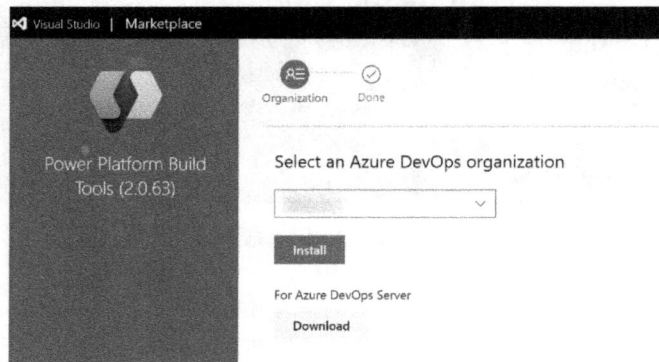

Figure 9.6 – Power Platform Build Tools install screen

> **Note**
> This action also needs elevated privileges in ADO. You might need your administrator to do or approve this step.

The Build Tools will be installed at an organization-wide level, and after this is done once, they can be used in as many projects as you'd like within your organization.

Service connections

To be able to really connect from ADO to our Dataverse environments, we have to add service connections. Those will then be used in our pipelines to connect to the correct environments.

You will need the following information to create a connection:

- **Dataverse URL**: This is the URL you use to access your Dataverse (for example, `https://<instancename>.crm4.dynamics.com`)
- **Tenant ID**: Can be found on the overview page of app registration
- **Client ID (or application ID)**: Can be found on the overview page of app registration
- **Client secret (or application secret)**: We created this during app registration

If all the prerequisites are met, you can follow these steps:

1. Go to **Project Settings** of your ADO project.
2. Select **Service Connections** (it's in the **Pipelines** group).
3. Click on **New service connection**:

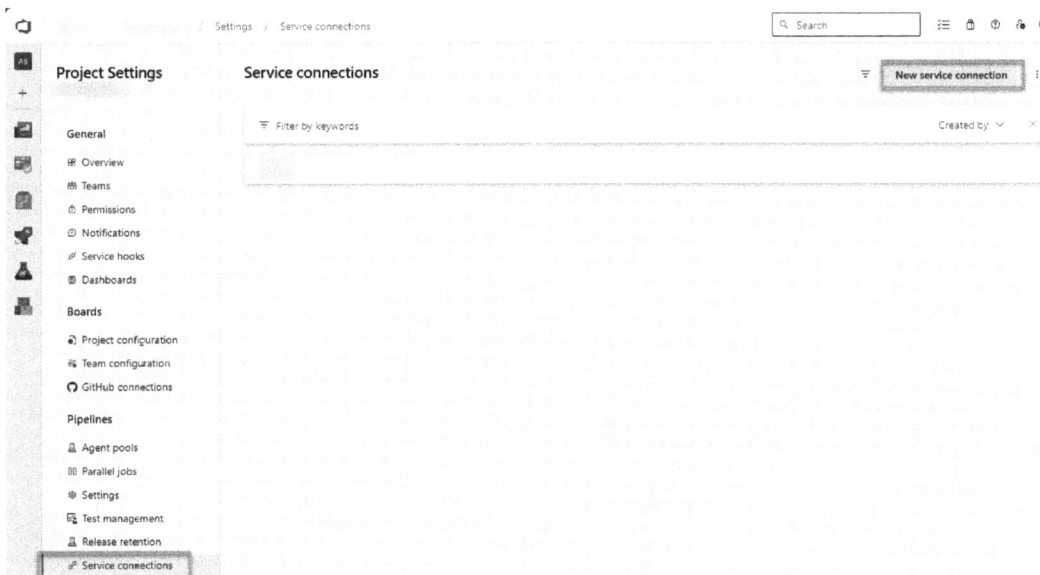

Figure 9.7 – Navigate to Service connections

4. Select **Power Platform** from the list:

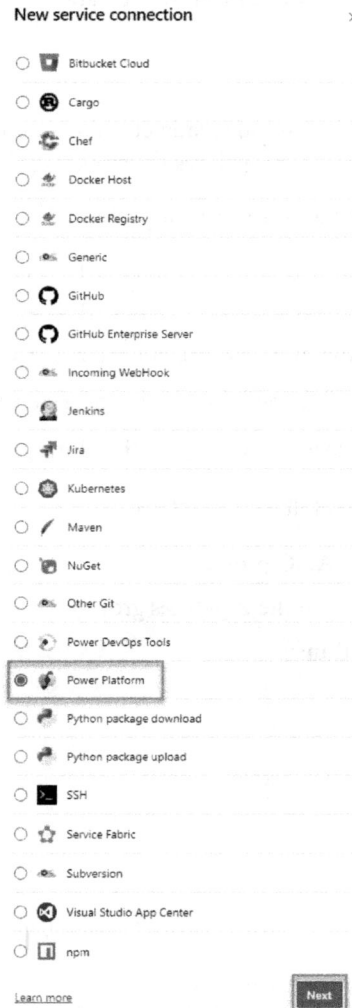

Figure 9.8 – Selecting the service connection type

5. Fill in all the needed information, such as the following:

- Server URL

- Tenant ID

- Application ID

- Client secret of the application ID

6. Give the connection a name and save it. That name is what you need to add to the variable files when it comes to the connections.

7. Make sure you have checked the checkbox beside **Grant access permission to all pipelines**.

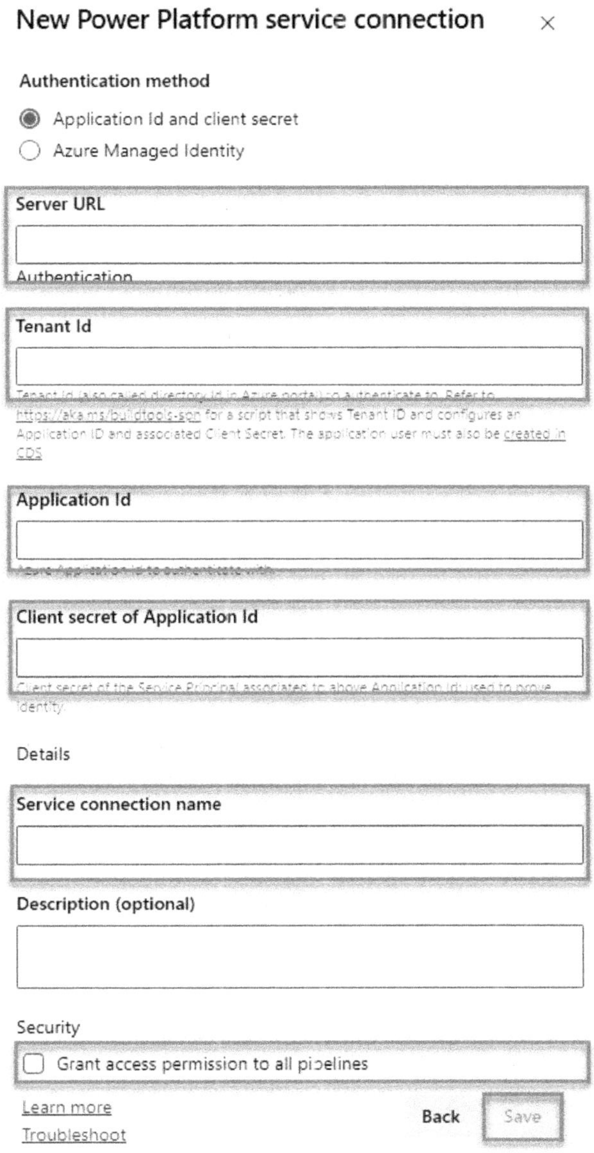

Figure 9.9 – Configuring the service connection

8. Click **Save**.

This needs to be done for every environment you'd like to use (usually DEV, Test, and Prod), even if the AppReg you are using is the same.

> **Note**
> Service connections are project-wide and can be reused from different repositories or pipelines within that given project.

Environments

The pipeline uses the concept of ADO environments. The idea is to make change tracking easier and add the possibility for pre-pipeline execution approvals. In ADO. under **Pipelines**, there's an **Environments** menu item. Go there, then click the **New environment** button:

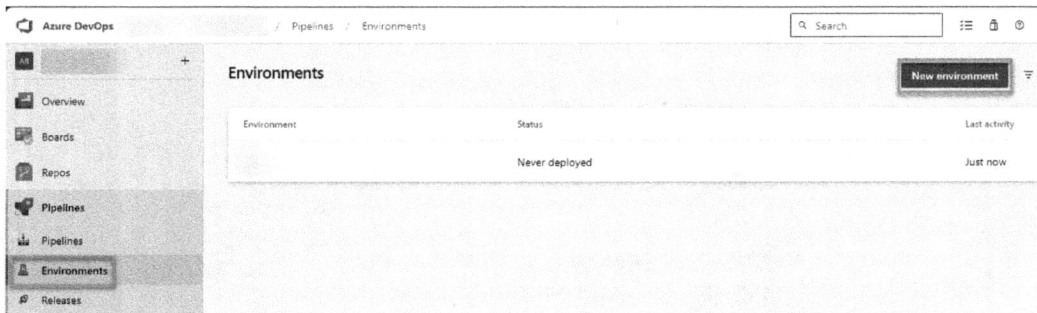

Figure 9.10 – Navigating to ADO environments

Here, we should create at least two environments: Test and Prod. Neither of those will have any resources since we neither select **Kubernetes** nor **Virtual machines**:

Figure 9.11 – Creating a new ADO environment

Depending on how you set up your system, you might need more environments. Let's go back to the previous example of different departures. In that case, we might have `Finance Test`, `Finance Prod`, `HR Test`, and `HR Prod`.

We only need Test and Prod since ADO environments are only used for target environments and not source environments.

The first time a pipeline tries to access an environment, it needs general approval. When the pipeline comes to the stage where it uses a new environment, it will ask for approval in the UI.

Note

ADO environments are project-wide and can be reused from different repositories or pipelines within that given project.

Please be aware that there is no relationship between service connections and ADO environments. If you deploy to the Test ADO environment but use the Production Service connection, it will still end up in your Dataverse production environment.

Approvals

We should add a pre-pipeline execution approval for your Prod environment. The pipeline is set up such that it will only deploy to Prod if the configured approval is met. If you do not add this, the pipeline will immediately start deploying to Prod as soon as the deployment to Test is ready.

To add an approval, follow these steps.

Open the environment in ADO and select **Approvals and checks** from the three-dot menu in the upper-right corner.

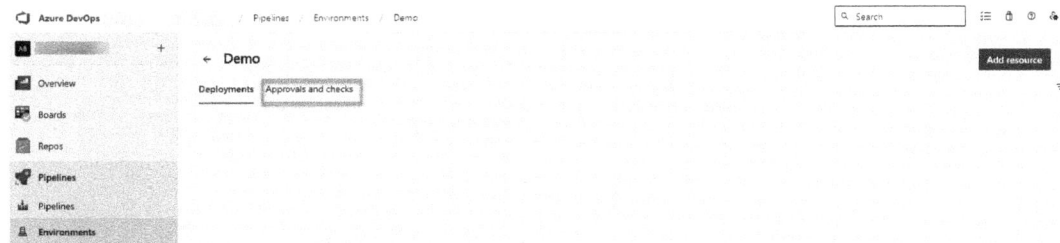

Figure 9.12 – Navigate to ADO environment approvals

On the next page, select + in the upper-right corner. In the panel that appears, select **Approvals** and click **Next**.

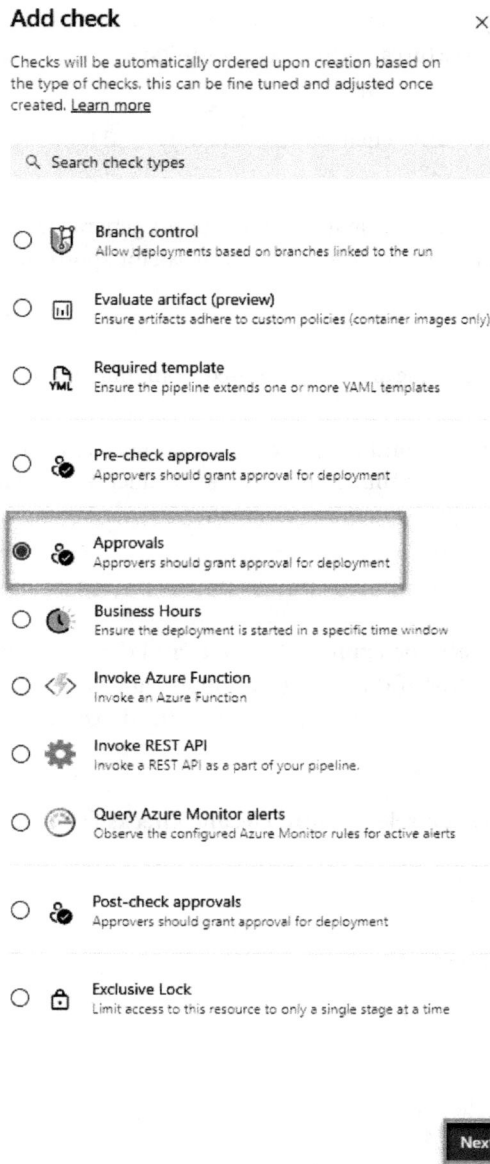

Add check   ✕

Checks will be automatically ordered upon creation based on
the type of checks, this can be fine tuned and adjusted once
created. Learn more

🔍 Search check types

○ 🛡️ **Branch control**
          Allow deployments based on branches linked to the run

○ 📊 **Evaluate artifact (preview)**
          Ensure artifacts adhere to custom policies (container images only)

○ 📄 **Required template**
   YML Ensure the pipeline extends one or more YAML templates

○ 👥 **Pre-check approvals**
          Approvers should grant approval for deployment

◉ 👥 **Approvals**
          Approvers should grant approval for deployment

○ 🕐 **Business Hours**
          Ensure the deployment is started in a specific time window

○ ⟨⟩ **Invoke Azure Function**
          Invoke an Azure Function

○ ⚙️ **Invoke REST API**
          Invoke a REST API as a part of your pipeline.

○ 🕤 **Query Azure Monitor alerts**
          Observe the configured Azure Monitor rules for active alerts

○ 👥 **Post-check approvals**
          Approvers should grant approval for deployment

○ 🔒 **Exclusive Lock**
          Limit access to this resource to only a single stage at a time

                                                    Next

Figure 9.13 – Selecting ADO environment approval

Next, select the people that are allowed to approve the deployment. In the **Advanced** configuration, select whether all of the approvers need to approve or just a certain amount (this is only visible if you have more than one approver added). Also, add whether an approver is allowed to approve their own run.

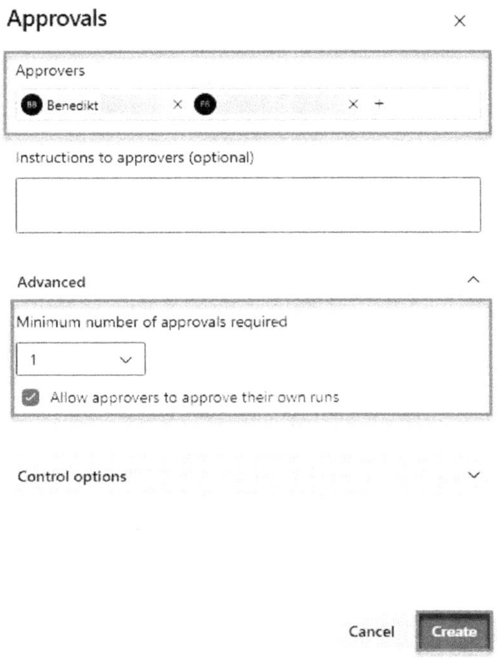

Figure 9.14 – Creating approvals

We usually select that when only one approver is needed and they can approve their own run.

> **Note**
>
> This configuration depends very much on which requirements your organization has. You definitely want to have at least one approver because of the mentioned reasons. In some organizations, there is a dedicated team responsible for all production environments regardless of which Application or technology is used. In those cases, you could add those users as approvers for the production environment. With this setup, the dedicated team would have to approve your deployment before you can deploy to Production.

Repo permissions

The export pipeline will push changes to our repository. To be able to do that, it needs certain permissions (Contribute should be set to allow) on the repository in question. To grant those, we have to do the following:

1. Open **Project Settings**.
2. Navigate to **Repositories**.

3. Select the repository in question.

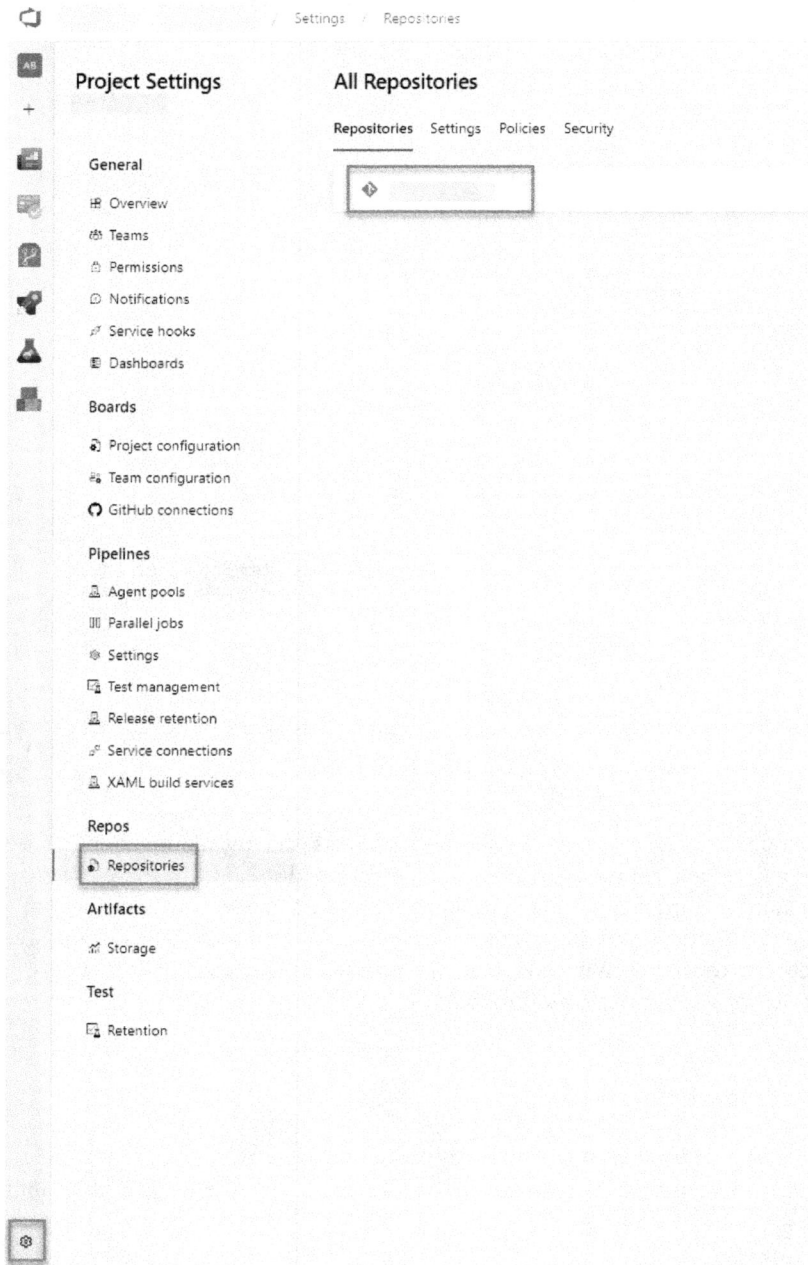

Figure 9.15 – Navigate to repository permissions

4. Open the **Security** tab.

5. Make sure that both the user called **Project Collection Build Service** and **<Project Name>
Build Service** have **Allow** for the **Contribute** permission.

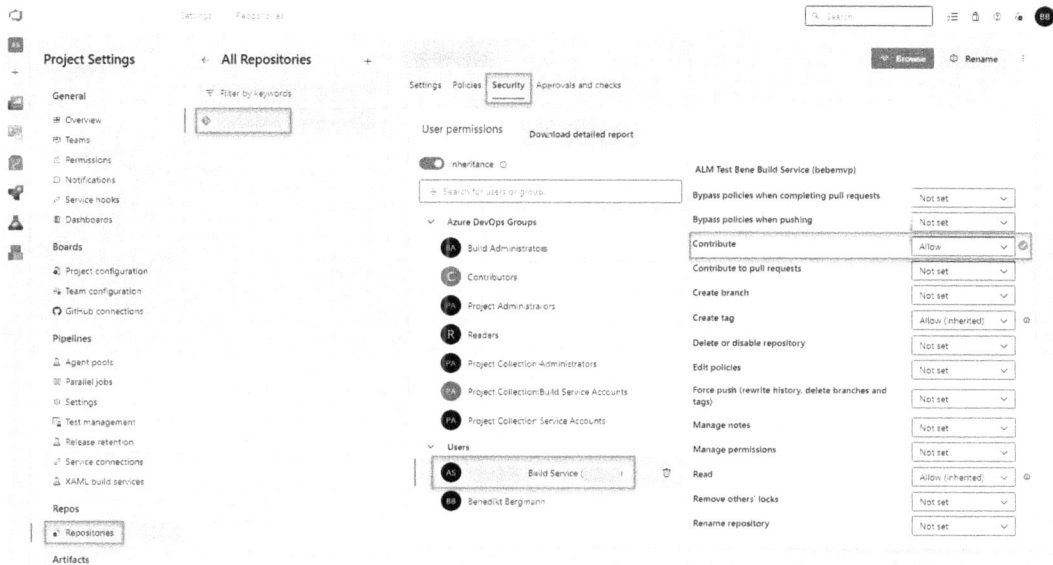

Figure 9.16 – Configuring repo permissions

The mentioned permissions are repository-wide and will be reused from different pipelines within
that given repository.

The following section will describe what is to be done in GitHub to be able to run pipelines in Dataverse.

Setting up GitHub

The setup for GitHub is simpler as compared to ADO. For example, the official Power Platform Actions
do not need to be installed. It is possible to just use them:

```
https://github.com/marketplace/actions/powerplatform-actions
```

Let's take a look at what we have to configure.

Environments

We have to add new environments for all the Dataverse environments we will use in the workflow.

To do that, navigate to the repository settings, then to **Environments**, and click on **New environment**.

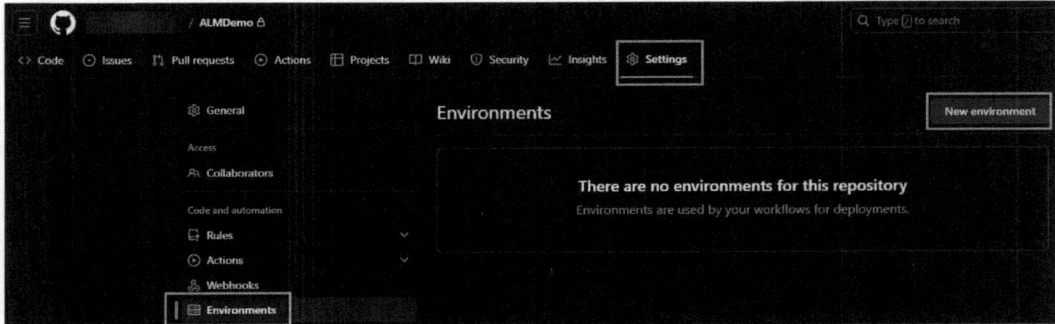

Figure 9.17 – Adding a new environment to GitHub

GitHub does not recognize any concept comparable to service connections in ADO. Therefore, we have to add **URL**, **CLIENTID**, and **CLIENTSECRET** to the environment as variables. On the **Environment configuration** page, we can select **Add secret** or **Add variable** to add either of them. A secret has the advantage that its value is never written to logs and should therefore be used for secrets. If either of the **Add** buttons is selected, we can specify a name and value. A common naming convention is to only use uppercase letters. Later in this book, we will explore how to use them.

You can read more about variables at `https://docs.github.com/en/actions/learn-github-actions/variables#creating-configuration-variables-for-an-environment`.

Approvals can also be added to the environment configuration. The possibilities for approvals are more basic than they are in ADO. Here, we only can specify up to six approvers and whether they can self-review the pull request.

You can read more about approvals/reviewers at `https://docs.github.com/en/repositories/configuring-branches-and-merges-in-your-repository/managing-protected-branches/about-protected-branches#require-pull-request-reviews-before-merging`.

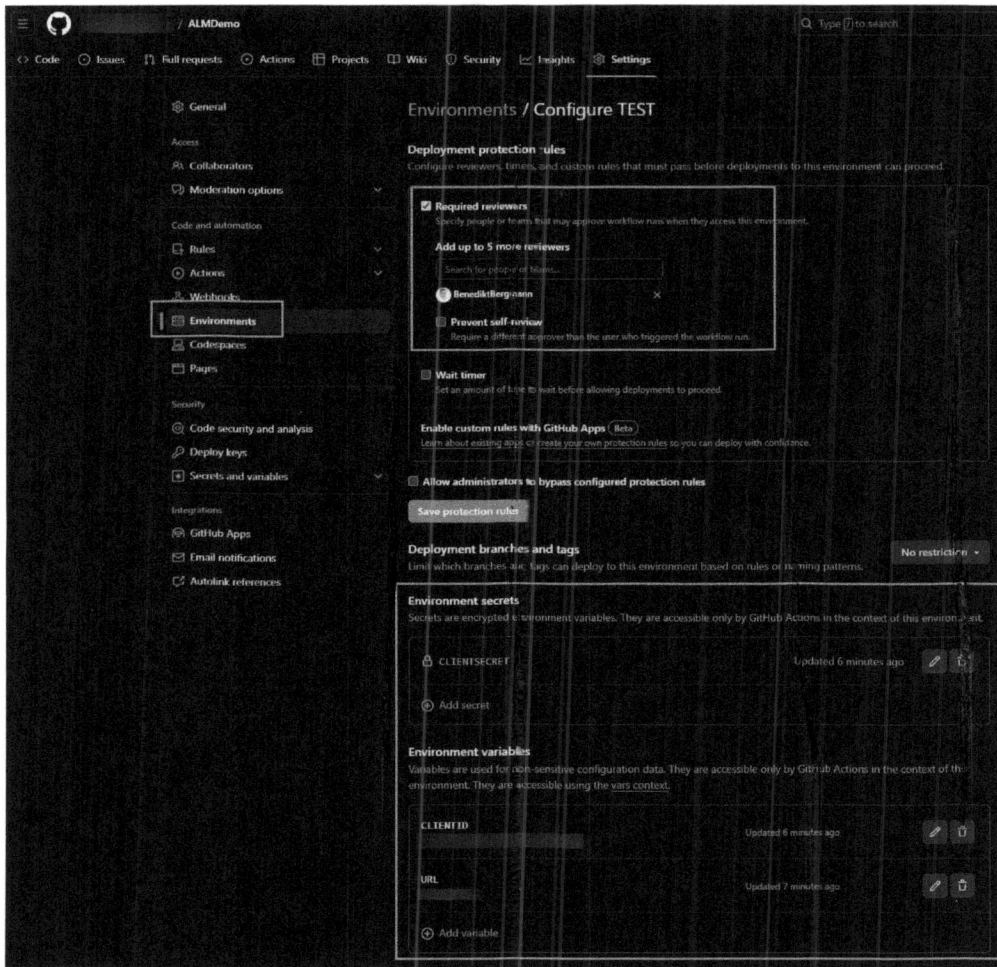

Figure 9.18 – Configuring the GitHub environment

> **Note**
>
> Required reviewers are not available for private repositories if you are on a free GitHub plan. This means that the upper box on the screenshot might not appear for you.

Repo Permission

Like in ADO, we must give the workflow runner permission to make changes to the repository. Otherwise, the step of committing changes to the repository will fail.

Under **Settings | Actions | General**, **Workflow permissions** need to be changed to **Read and write permissions**:

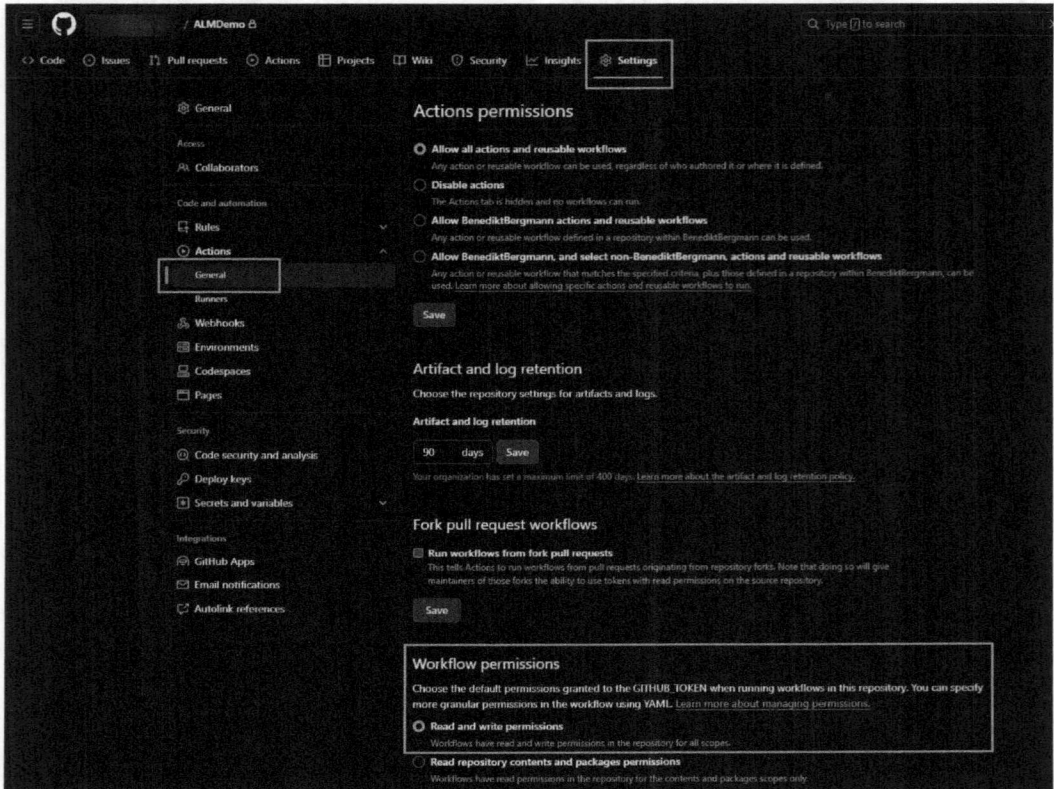

Figure 9.19 – Setting read and write permissions for GitHub

In the following section, we will briefly take a look at how to configure the ALM Accelerator.

Configuring the ALM Accelerator

The setup of the ALM Accelerator is rather complex compared to the setup of all other options you have (such as GitHub, ADO, or PPP).

It also changes slightly every now and then. Also, there is very good documentation on it (see the *Further reading* section). Therefore, we will not go into much detail. In general, the installation contains the following:

- **Setting up app registrations**: The ALM accelerator also needs app registrations. The setup for these is described earlier in this chapter.

- **Cloning YAML files to your ADO**: We have to clone the Microsoft YAML files so that our ALM Accelerator installation can run the same pipelines in our ADO.

- **Setting up permissions**: As "normal" pipelines, the pipelines of the Accelerator also need certain permissions.

- **Creating service connections**: The pipelines of the Accelerator need access to our Dataverse instance and therefore service connections within ADO.

- **Installing several solutions**: To use the full potential of the Accelerator, we need to install some Power Platform solutions in a dedicated environment. After that, we can configure and use the Accelerator through Power Apps.

- **Configuring those solutions**: To be able to run the first pipelines with the Accelerator, some configurations have to be done.

You can read more about it in Microsoft Learn at `https://learn.microsoft.com/en-us/power-platform/guidance/alm-accelerator/overview`.

Configuring PPP

Before we can start with PPP or pipelines in Power Platform, there are a few things (for example, installing solutions and connecting environments) to set up.

Microsoft recommends having a host environment whose only purpose is to configure all pipelines in Power Platform. This is because to use PPP, a solution needs to be installed. This solution will add dependencies on several OOB Tables. To not disrupt your implementation, the host environment should be, as mentioned, a separate environment.

To install the mentioned solutions go to PPAC, select the environment, and open **Dynamics 365 apps** under **Resources**.

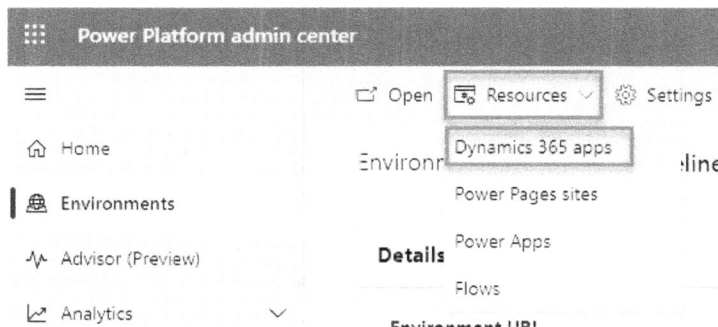

Figure 9.20 – Navigating to Dynamics 365 apps

Under Dynamics 365 apps, select **Install app**. A new side pane appears on the right-hand side of the page. Here, search for **Power Platform Pipelines**, select **Next**, agree to the terms, and press **Install**.

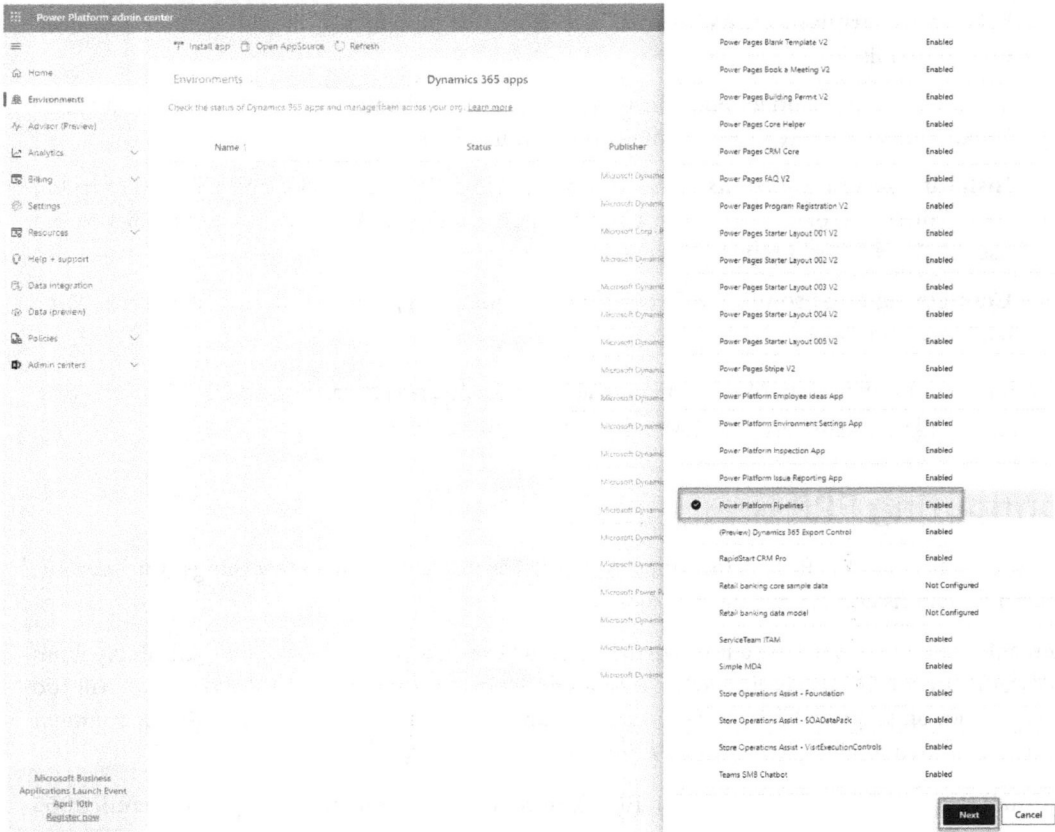

Figure 9.21 – Installing PPP

> **Note**
>
> PPP requires the involved target environments to be either of type Development, Sandbox, or production. They have to be managed environments. The next section will explain how to activate an environment to be managed.

When the **Power Platform Pipeline** app is installed, you will find a new **Model-Driven App (MDA)**, called **Deployment Pipeline Configuration**, in your **Host** environment.

In the mentioned app, we have to set up all environments involved in the future pipeline. Most of the time, this is our Development, Test, and Production environment. To do so, navigate to **Environments** in the MDA and add a new environment. Both the **Environment Type** and the **Environment ID** are important. The **Environment Type** determines whether the environment you are adding is a development environment (source) or a target environment (`Test`, `Prod`, or any other).

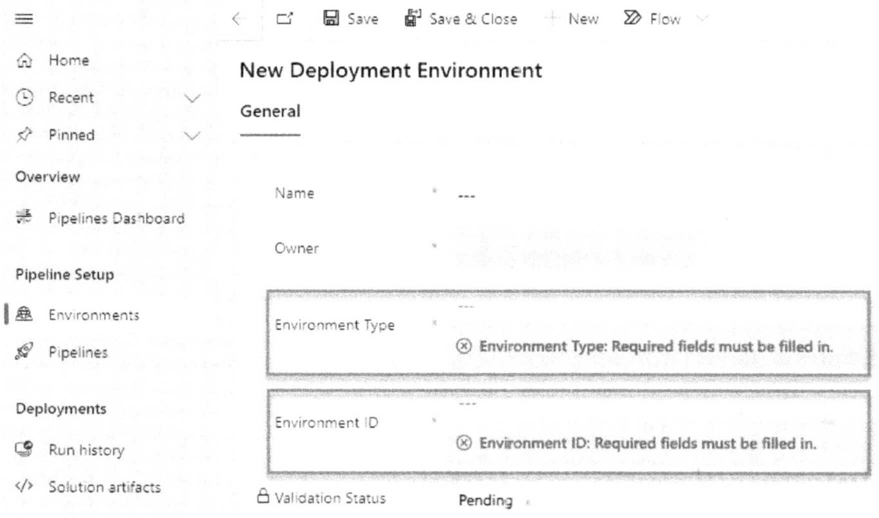

Figure 9.22 – Adding an environment to PPP

The **Environment ID** can be found on the **Overview** page of an environment in the PPAC.

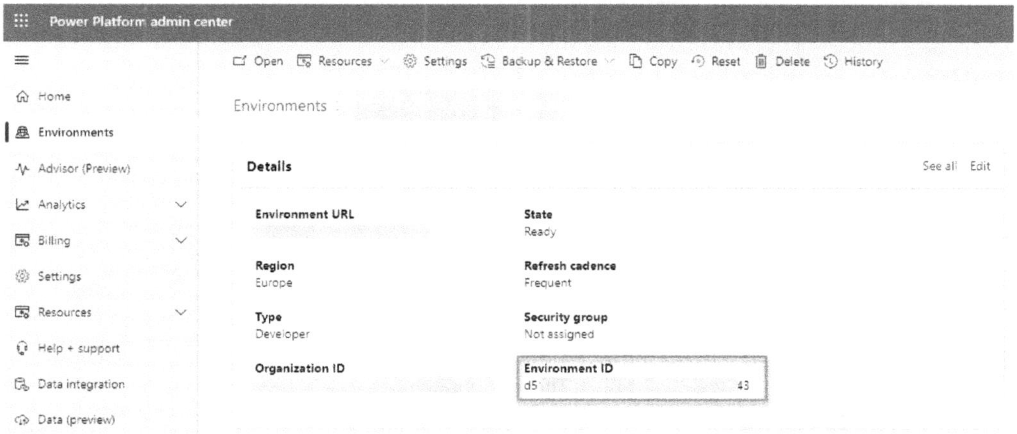

Figure 9.23 – Getting the environment ID

When a new environment is added to the PPP config MDA, a background process kicks off to check whether the environment exists, and whether the user has access to it. When this check is successful, the validation status will change to **Success**.

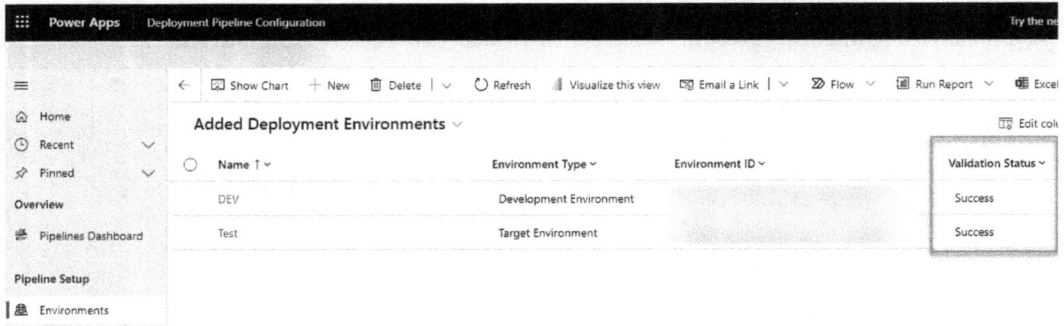

Figure 9.24 – Added environments to the configuration MDA

Since the PPP requires all target environments to be managed environments, the next section will describe how we can switch an environment to be managed.

Configuring managed environments

Activating an environment to be a managed environment is an easy task.

Navigate to the PPAC, select the environment in the list of environments, and press **Enable Managed Environments**.

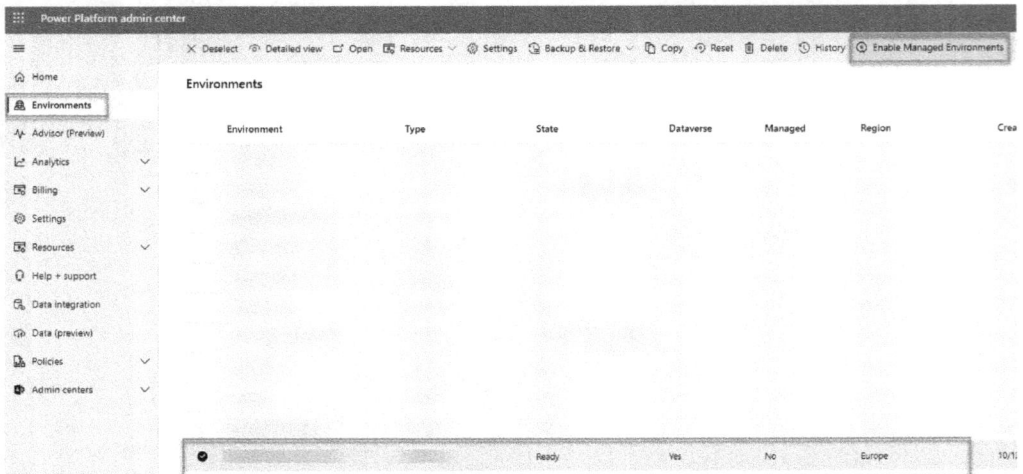

Figure 9.25 – Activating an environment to be managed

On the upcoming side pane, select the configurations you'd like to and confirm by pressing **Enable**. There are already quite a few possibilities (such as **Solution Checker enforcement** or **Usage insights**) and Microsoft is adding more and more features to managed environments at the moment.

Depending on the Dataverse version your environment is running on, you can switch back and forth between the environment being managed and not. For older versions, it is not possible to deactivate it. Only through a support ticket can you find out which version the Dataverse is.

> **Note**
>
> If an environment is activated to become a managed environment, every user using the environment in question is required to have a premium license.

The next section will describe a potential folder structure. This is needed since the pipelines we build require a certain folder structure. This can be altered to your needs.

Folder structure

This section will give you an overview of how a folder structure in your repository might look to help you structure your Power Platform development. Note that this is not the only possibility; there are many right answers to this question. This section only shows one possible solution. The setup is also intended for having a mono repo. This means one repository holding all the code for your implementation. If you use separate repos for separate parts of your project (Azure Functions in one repo, Dataverse plugins in another, configuration in another still, and so on), this structure might not suit your needs. In addition, the pipelines will get more complex.

Root folder

In the `root` folder, you could have the following

- `Development`: This folder would contain all your actual development. For example, it might contain plugins, TypeScript, or surrounding applications such as Azure Functions or APIs.

- `PipelineDefinition`: As the name suggests, this folder would contain your pipeline definitions (YAML files). However, it also could contain your Bicep files to create Azure components in an automated manner along with your Dataverse deployment.

- `PowerPlatform`: In the `PowerPlatform` folder, you can store everything that is related to Power Platform. This can be the unpacked versions of your exported Solutions, settings files that are used for import, third-party solutions (for example, from ISVs), data export configurations, or many other components.

Development

Let's dive a bit deeper into the `Development` folder.

It could contain the following:

- `Back-End`: The `Back-End` folder could, as the name suggests, contain all your development for backend components such as plugins, workflow custom actions, Azure Functions, or APIs. Basically, it could contain everything that is written in C#.

  It would also contain shared projects such as EarlyBound definitions or your internal Plugin Base class. Those projects should then be reused in your different main projects.

- `Front-End`: In the `Front-End` folder, everything related to the frontend should be stored. This could be your client-side scripts developed with TypeScript, web resources in general, Power Platform Component Framework Controls, or UI tests.

Screenshot

The following screenshot is an illustration of the described folder structure.

Figure 9.26 – The described folder structure

Summary

In this chapter, we covered a range of topics related to GitHub, Azure DevOps (ADO), managed environments, the ALM Accelerator, and general folder structuring. This included setting up app registration, configuring permissions in ADO, and managing environments within ADO. We also discussed establishing service connections in ADO, managing environments in GitHub, and configuring permissions in GitHub. Additionally, we explored how to activate an environment to be managed and how to set up the ALM Accelerator.

The next chapter will go into depth on how to bring all the learned knowledge into play and create source code-centric YAML pipelines.

Questions

1. Which two subfolders does the root have?

   A. Development

   B. Frontend

   C. PipelineDefinition

   D. Backend

2. Which repository permission do we have to set on the Build service user?

   A. Contribute Allow

   B. Edit policy allow

   C. Create Branch allow

3. Which scope do ADO environments have?

   A. Organization-wide

   B. Project-wide

   C. Repository-wide

Further reading

If you want to set up the ALM Accelerator, you should follow the Microsoft documentation on it. This can be accessed at `https://learn.microsoft.com/en-us/power-platform/guidance/alm-accelerator/setup-admin-tasks`.

On the following docs site, you can read more about the functionalities of managed environments: `https://learn.microsoft.com/en-us/power-platform/admin/managed-environment-overview`.

10
Pipelines

In this chapter, we will learn how to implement an easier version of the environment-centric approach using **Power Platform Pipelines** (**PPP**), as well as the source code-centric approach using **Azure DevOps** (**ADO**) and **GitHub Actions**.

In addition, we will include a brief overview of what **Yet Another Markup Language** (**YAML**) is and how we write it.

After that, the part about the PPP will also include some current pitfalls and disadvantages, as well as advantages, compared to the *full-blown* systems.

The YAML syntax between ADO and **GitHub** (**GH**) is slightly different. Therefore, we will provide snippets for both platforms in the following chapters.

This chapter includes the following topics:

- What is YAML
- Source code-centric approach
- General setup of a new pipeline
- Exporting pipeline
- Building pipeline
- Release pipeline
- Running a pipeline
- Environment-centric approach

Technical requirements

To be able to execute all the steps described in this chapter, you will need the following technical requirements:

- An ADO or GH project
- PAC CLI installed
- At least two Dataverse environments
- App registration added to the Dataverse environments

Please be aware that all the .yaml files we create in this book can be found in our complimentary GH repository. We do this to make it easier for you since you don't have to type everything yourself: https://github.com/PacktPublishing/Application-Lifecycle-Management-on-Microsoft-Power-Platform

What is YAML

Since pipelines, both in ADO and GitHub Actions, use YAML to define what should be done, we will take a brief look at what YAML is and how the syntax should be written.

YAML can also stand for **YAML Ain't Markup Language**.

It is a human-readable data serialization language and is widely used for configuration files. The syntax is intentionally minimal and uses a Python-style indentation to show nested objects.

Let's take the following YAML snippet as an example:

```
stages:
- stage: DeployTest
  displayName: 'Deploy to Test'
  jobs:
  - job: deployTest
    pool:
      vmImage: 'windows-latest'
    steps:
    - checkout: self
      clean: true
      submodules: false
      displayName: "Checkout repo"
- stage: DeployProd
  displayName: 'Deploy to Prod'
  jobs:
  - job: deployProd
```

```
pool:
  vmImage: 'windows-latest'
steps:
- checkout: self
  clean: true
  submodules: false
  displayName: "Checkout repo"
```

It shows part of a release pipeline written in YAML. First, the stages are defined (`DeployTest` and `DeployProd`). Within every stage, there are jobs with steps. The indentation is important to specify where each part belongs and what is grouped together.

Source code-centric approach

In *Chapter 7*, we explained what the source code-centric approach is. Since we will, in this chapter, create pipelines, both for ADO and GitHub Actions, to implement this approach, we will do a quick summarization of it to bring it back to your mind.

The source code-centric approach needs three different pipelines with the following steps:

- **Export Pipeline**:

    I.   Install tooling.

    II.   Publish customizations.

    III.   Export as unmanaged.

    IV.   Export as managed.

    V.   Unpack.

    VI.   Commit.

- **Build Pipeline**:

    I.   Install tooling.

    II.   Pack the solution from the repository.

    III.   Publish artifacts.

- **Release Pipeline**:

    I.   Install tooling.

    II.   Import solution.

    III.   (Apply upgrade).

In the following sections, we will see how those pipelines are actually implemented in ADO pipelines as well as GitHub Action workflows.

> **Note**
>
> All the pipelines and workflows we build in this chapter are very rudimentary. They do the job they should but don't do anything advanced. In the next chapter, we will look into ways of adding more functionality to make the pipelines more sophisticated.

General setup of a new pipeline

First of all, we have to create a new pipeline. This step has to be done for all the three pipelines we need.

After the pipelines/workflows have been created in either ADO or GH, the steps are specific to every pipeline. We will go into depth regarding them in the sections right after the general setup.

ADO

We have to navigate to the ADO project where the pipeline should run. There, we open the **Pipeline** menu and select the **New pipeline** button on the upper-right corner of the page.

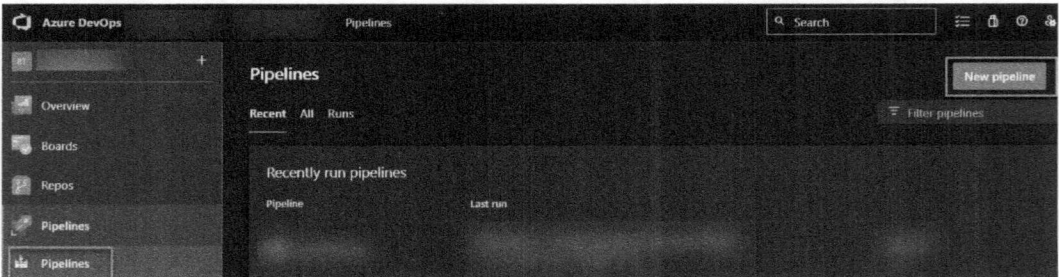

Figure 10.1: Creating a new pipeline in ADO

> **Note**
>
> If there isn't a pipeline in the project yet, there will be a **New Pipeline** button in the middle of the page.

On the following screen, we select **Azure Repos Git**:

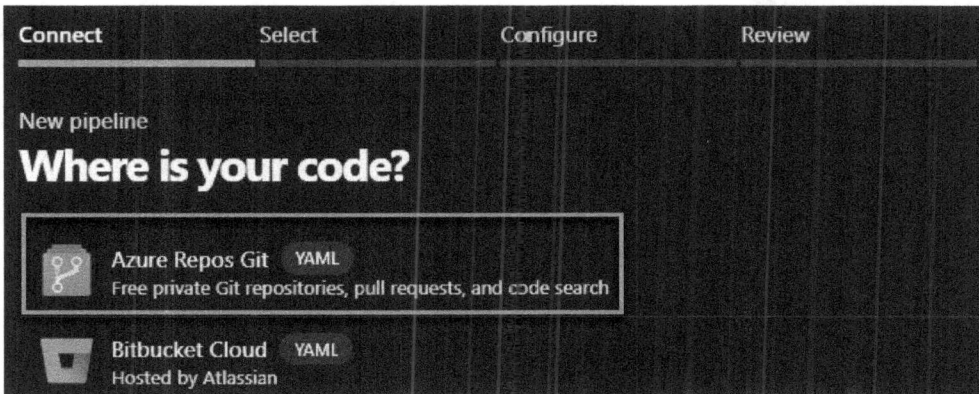

Figure 10.2 Selecting Azure Repos Git as the source

Thereafter, we select the correct repository to use. Every project has at least one repository that has the same name as the project. Often, there is only one per project, but sometimes there are more. Either way, we have to select the correct one.

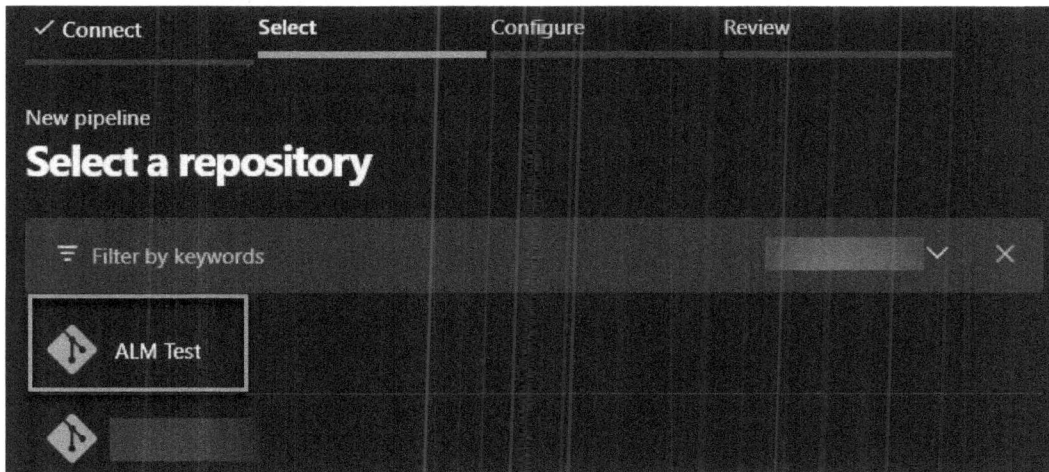

Figure 10.3: Selecting a repository

The next step is to select either a template or an existing YAML file as the source or to create a new pipeline from scratch (which is what we will do).

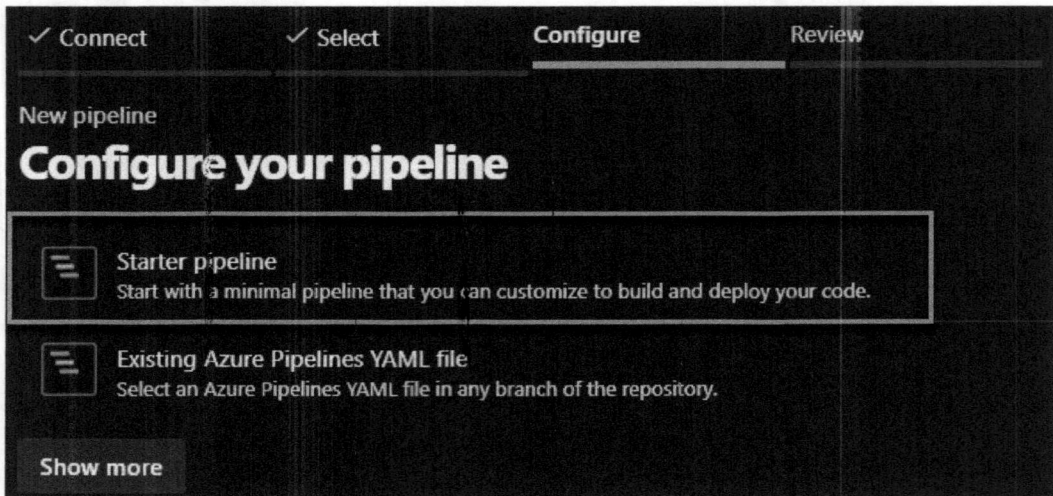

Figure 10.4: Selecting a template

> **Note**
> Depending on which programming languages you have in your repository, the shown list of templates might vary.

To make the learning easier, the pipelines we build will not be automated. They all need to be run manually. In a perfect ALM world, there might be good reasons to run a certain pipeline whenever a trigger occurs (this could be, for example, if a previous pipeline runs successfully, a commit to the repo has been made, or something different). This highly depends on the project and the needs it has.

For now, we replace the content of the file with the following:

```
trigger: none
pr: none

pool:
  vmImage: 'windows-latest'

steps:
```

With that, we have defined that there should not be an automated trigger and that the pool of agents should be used with the latest Windows version.

We also have to change the name of the underlying .yaml file. I tend to use something that says more than the default `azure-pipeline-1.yml` file, for example, `export.yml`, `build.yml`, and `release.yml`.

To change the name, simply click on the current name and change it.

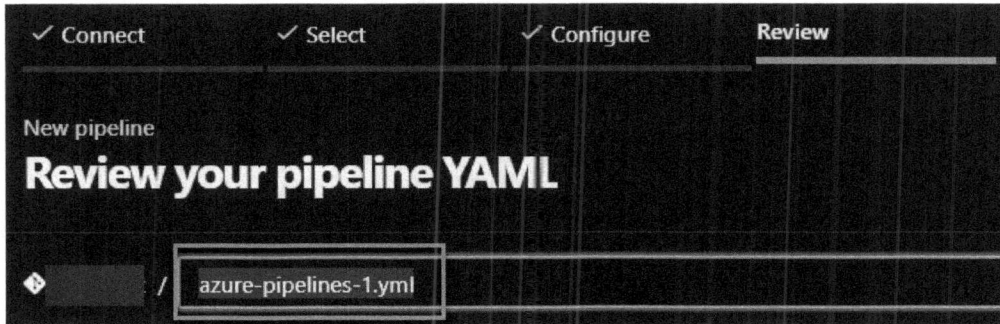

Figure 10.5: Changing the .yml file name

When the pipeline is saved, it will get the same name as the repository. That means that we have to manually change the name to something more useful such as Export from DEV, Build Solution, or Release depending on which pipeline we have created. To do this, go to the list of all pipelines.

> **Note**
>
> Pipelines that have not been run yet will not show under the default tab, **Recent**, of the **Pipelines** menu. To be able to see them, you have to switch to the **All** tab.
>
>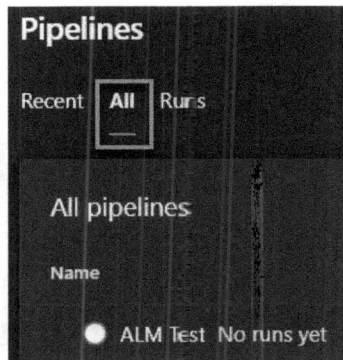
>
> Figure 10.6: Tab where not-run pipelines are shown

If you hover over the pipeline in question, three dots will appear on the right side. One of the options is **Rename/Move**. In the appearing popup, you can change the name.

Figure 10.7: Changing the name of a pipeline

In this section, we learned how to set up a pipeline in ADO. The next section will explain how to do that in GH.

GH

When it comes to GH, the initial creation per workflow looks a bit different. Within the GH project in question, we navigate to **Actions** and select the **New workflow** button.

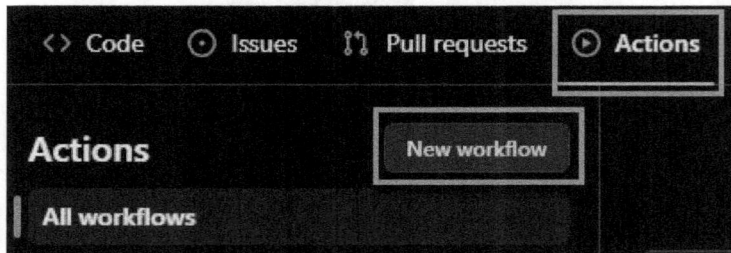

Figure 10.8: New workflow

On the following page, we select **Simple workflow** as the template.

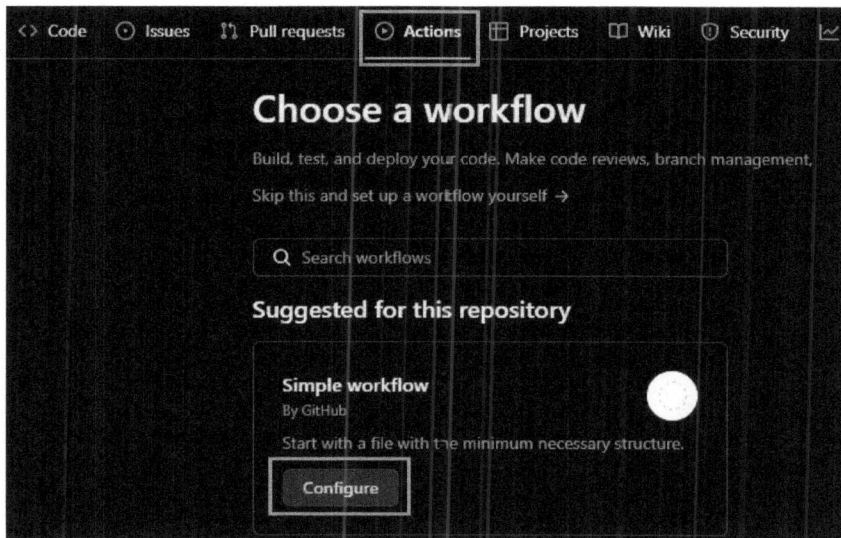

Figure 10.9: Selecting Simple workflow

We replace the whole file with the following content:

```
name: Export From DEV
on:
  workflow_dispatch:

jobs:
  Export:
    runs-on: windows-latest
    environment: DEV
    steps:
```

The name has to be altered depending on the workflow. The build pipeline will not have an environment; therefore, the build pipeline line could be deleted here. For the others, the environment has to be changed to whichever environment is targeted with the job in question.

One last thing to do is to give the underlying `.yml` file a descriptive name. This can be done by changing the name in the breadcrumb before doing the initial commit.

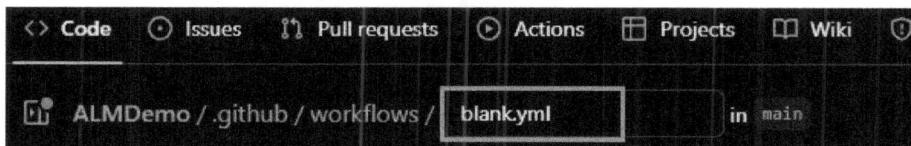

Figure 10.10: Renaming a GH workflow

In this section, we learned how to create a workflow in GH.

Now that we have created the basic setup of pipelines both for ADO and GH, it is important to mention that the steps described in the following sections need to be placed under "steps:" in the already created YAML.

Export pipeline

Let's start with the Export pipeline. It is the pipeline with the most steps and therefore the most complex one we will create in this chapter.

Checkout (GH only)

Before we install the tooling, GH needs an additional step to check out the repository.

The YAML is as follows:

```
- name: Checkout
  uses: actions/checkout@v4
```

Now we are ready to install the tooling.

Installing the tooling

One of the first steps we always have to add is to install the Power Platform tooling. This basically installs the Power Platform CLI onto the agent. When you start out with ADO, your pipelines will run on shared agents. This means that every run could end up on a different agent than the run before. Since those agents are shared between all ADO tenants and will be reset after every run, we can't be sure that the very specific tooling for Power Platform is already installed. Therefore, we have to make sure it is.

The YAML for ADO looks like the following:

```
- task: PowerPlatformToolInstaller@2
  displayName: 'Power Platform Tool Installer '
```

The YAML for GH looks like the following:

```
- name: Install Power Platform Tools
  uses: microsoft/powerplatform-actions/actions-install@v1
```

Publishing customizations

After we have installed the tooling, we have to publish all customizations in the development environment to be sure that all the changes are part of our export.

ADO

In the `PowerPlatformSPN` parameter, you need to provide the name of the service connection to the development environment that we created earlier:

```
- task: PowerPlatformPublishCustomizations@2
  displayName: Publish Customizations
  inputs:
    authenticationType: 'PowerPlatformSPN'
    PowerPlatformSPN: '<Name of the dev ADO Service connection>'
    AsyncOperation: true
    MaxAsyncWaitTime: '60'
```

GH

The `environment-url`, `app-id`, `client-secret`, and `tenant-id` parameters come from the environment variables we created earlier. The `solution-name` parameter has to be changed to the name of your solution:

```
- name: Publish Solution
  uses: microsoft/powerplatform-actions/publish-solution@v1
  with:
    environment-url: ${{ vars.URL }}
    app-id: ${{ vars.CLIENTID }}
    client-secret: ${{ secrets.CLIENTSECRET }}
    tenant-id: ${{ vars.TENANTID }}
    solution-name: <Solution Schemaname>
```

Exporting as unmanaged

The next step is to export the solution in question as unmanaged.

ADO

It is important to adjust the `PowerPlatformSPN`, `SolutionName`, and `SolutionOutputFile` parameters to your needs:

```
- task: PowerPlatformExportSolution@2
  displayName: Export Solution - Unmanaged
  inputs:
    authenticationType: 'PowerPlatformSPN'
    PowerPlatformSPN: '<Name of the dev ADO Service connection>'
    SolutionName: '<Solution Schemaname>'
```

```
SolutionOutputFile: '$(Build.ArtifactStagingDirectory)\\ <Solution
Schemaname>.zip'
AsyncOperation: true
MaxAsyncWaitTime: '60'
```

GH

The `solution-name` and `solution-output-file` parameters have to be changed to suit your needs:

```
- name: Export Solution - Unmanaged
  uses: microsoft/powerplatform-actions/export-solution@v1
  with:
    environment-url: ${{ vars.URL }}
    app-id: ${{ vars.CLIENTID }}
    client-secret: ${{ secrets.CLIENTSECRET }}
    tenant-id: ${{ vars.TENANTID }}
    solution-name: <Solution Schemaname>
    solution-output-file: "${{runner.temp}}/exported/<Solution
    Schemaname>.zip"
    managed: false
    run-asynchronously: true
    overwrite: true
```

Exporting as managed

We also need to export the solution as a managed solution.

ADO

As in the previous step, it is important to alter the `PowerPlatformSPN`, `SolutionName`, and `SolutionOutputFile` parameters to fit your needs:

```
- task: PowerPlatformExportSolution@2
  displayName: Export Solution - Managed
  inputs:
    authenticationType: 'PowerPlatformSPN'
    PowerPlatformSPN: '<Name of the dev ADO Service connection>'
    SolutionName: '<Solution Schemaname>'
    SolutionOutputFile: '$(Build.ArtifactStagingDirectory)\\ <Solution
    Schemaname>_managed.zip'
    Managed: true
    AsyncOperation: true
    MaxAsyncWaitTime: '60'
```

GH

The `solution-name` and `solution-output-file` parameters have to be changed to suit your needs:

```
- name: Export Solution - Unmanaged
  uses: microsoft/powerplatform-actions/export-solution@v1
  with:
      environment-url: ${{ vars.URL }}
      app-id: ${{ vars.CLIENTID }}
      client-secret: ${{ secrets.CLIENTSECRET }}
      tenant-id: ${{ vars.TENANTID }}
      solution-name: <Solution Schemaname>
      solution-output-file: "${{runner.temp}}/exported/<Solution
      Schemaname>_managed.zip"
      managed: true
      run-asynchronously: true
      overwrite: true
```

In this example, we only handle one solution. A real project usually has several solutions, as described earlier. In the next chapter, we will take a look at how this can be added to the pipeline.

Unpack

Now that we have exported our solution, both as managed and unmanaged, we can unpack it locally on the agent.

ADO

In this step, it is important to alter the `SolutionInputFile` and `SolutionTargetFolder` parameters to include the name of your solution:

```
- task: PowerPlatformUnpackSolution@2
  displayName: Unpack unmanaged Solution
  inputs:
    SolutionInputFile: '$(Build.ArtifactStagingDirectory)\\ <Solution
    Schemaname>.zip'
    SolutionTargetFolder: '$(build.sourcesdirectory)\PowerPlatform\
    Solutions\<Solution Schemaname>'
    SolutionType: 'Both'
```

GH

The `solution-file` and `solution-folder` parameters need to be changed to fit your project's needs:

```
- name: Unpack Solution
  uses: microsoft/powerplatform-actions/unpack-solution@v1
  with:
    solution-file: '${{runner.temp}}/exported/ <Solution Schemaname>.
    zip'
    solution-folder: 'PowerPlatform/Solutions/ <Solution Schemaname>'
    solution-type: 'Both'
    overwrite-files: true
```

> **Note**
>
> The unpack with `"Both"` only works when both the unmanaged and managed solution are in the same folder and the managed solution has the same name suffixed by _managed.

Commit

The last step of this pipeline is to commit the changes we did with the unpacking step to the repository.

ADO

The following code shows how to execute the commit in the ADO pipeline:

```
- task: CmdLine@2
  displayName: Commit to Repo
  inputs:
    script: |
      git config user.email $(Build.RequestedForEmail)
      git config user.name "$(Build.RequestedFor)"
      git -c http.extraheader="AUTHORIZATION: bearer $(System.
      AccessToken)" fetch
      git -c http.extraheader="AUTHORIZATION: bearer $(System.
      AccessToken)" checkout main
      git add --all
      git commit -m "$(Build.DefinitionName) $(Build.BuildNumber)"
      git -c http.extraheader="AUTHORIZATION: bearer $(System.
      AccessToken)" pull origin main
      git -c http.extraheader="AUTHORIZATION: bearer $(System.
      AccessToken)" push origin main
```

GH

You need to change the script to commit the correct path, write a correct commit message, and use the correct branch:

```
- name: Commit Changes to Repo
  run: |
    git config user.email "<>"
    git config user.name "${{github.triggering_actor}}"
    git add 'PowerPlatform/Solutions/ <Solution Schemaname>' --all
    git commit -m " <Solution Schemaname> Changes"
    git pull origin main
    git push origin main
```

Build pipeline

The next pipeline we have to create is the build pipeline. It only has the following four steps:

1.   Checkout (GH only)

2.   Install tooling

3.   Pack solution

4.   Publish artifact

Checkout (GH only)

Before we install the tooling, GH needs an additional step to checkout the repository.

The YAML is as follows:

```
- name: Checkout
  uses: actions/checkout@v4
```

Installing tooling

This step is exactly the same as it was for the export pipeline.

The ADO YAML is as follows:

```
- task: PowerPlatformToolInstaller@2
  displayName: 'Power Platform Tool Installer '
```

The GH YAML is as follows:

```
- name: Install Power Platform Tools
  uses: microsoft/powerplatform-actions/actions-install@v1
```

Packing the solution

As the second step, we have to pack our managed solution from source control. It is important to change the parameters to handle your solution as well as the folder structure you have.

ADO

The following YAML shows how to pack a solution in ADO. As always, you have to replace `<Solution Schemaname>` with the actual name of your solution:

```
- task: PowerPlatformPackSolution@2
  inputs:
    SolutionSourceFolder: '$(build.sourcesdirectory)\PowerPlatform\
    Solutions\<Solution Schemaname>'
    SolutionOutputFile: '$(Build.ArtifactStagingDirectory)\\ <Solution
    Schemaname>_managed.zip'
    SolutionType: 'Managed'
```

GH

The following snippet shows how to pack a solution using GH. As always, you have to replace `<Solution Schemaname>` with the actual name of your solution:

```
- name: Pack Solution
  uses: microsoft/powerplatform-actions/pack-solution@v1
  with:
    solution-file: '${{runner.temp}}/packed/ <Solution Schemaname>_
    Managed.zip'
    solution-folder: 'PowerPlatform/Solutions/ <Solution Schemaname>'
    solution-type: 'Managed'
    overwrite-files: true
```

Publishing the artifact

The last step of this pipeline is to publish the packed solution file as an artifact so that the release pipeline can use it.

ADO

The `artifact` parameter can be anything you would like. It is just important to use the same name in the release pipeline:

```
- task: PublishPipelineArtifact@1
  displayName: Publish Artifacts
  inputs:
    targetPath: '$(Build.ArtifactStagingDirectory)'
```

```
      artifact: 'drop'
      publishLocation: 'pipeline'
```

GH

The name parameter could be anything you would like. It is important to use the same name in the release pipeline:

```
- name: Publish Artifacts
  uses: actions/upload-artifact@v4
  with:
    name: drop
    path: '${{runner.temp}}/packed/**'
    overwrite: true
```

Release pipeline

The last pipeline is the release pipeline. Since the goal is to deploy with one pipeline run to different environments, this pipeline will be a bit more complex than the other two. We have to use different stages for different environments.

ADO

We also have to configure the build pipeline as a resource so that we can reuse the created solution file artifact.

To achieve this, we replace the whole file with the following YAML:

```
trigger: none
pr: none

resources:
  pipelines:
    - pipeline: buildPipeline
      source: 'Build Solution'
      trigger: none

stages:
```

In the resources part, it is important that the value of the source parameter exactly matches the name of your build pipeline.

After that, there is a setup per stage. Usually, we at least have **Test** and **Production**, but there could be other environments in between or after that. For example, a UAT or Hotfix environment could be used.

Here is the YAML for the Test stage as an example. This can be copied and altered for every stage or environment you would like to release/deploy to:

```
- stage: DeployTest
  displayName: 'Deploy to Test'
  jobs:
  - job: deployTest
    pool:
      vmImage: 'windows-latest'
    steps:
```

Now, we can add the steps the stage needs. Those will be the same for every stage, with the only exception that the `PowerPlatformSPN` parameter will have different service connections configured.

GH

When it comes to GH, we only change the `on:` configuration to automatically start the release workflow when the build workflow is done. This is the easiest way to access the same build artifact from every job or stage in the workflow. Replace *lines 6* and *7* with the following:

```
on:
  workflow_run:
      workflows: ['Build Solution']
      types: [completed]
      branches:
          - 'main'
```

As described earlier, the first five lines after `jobs:` repeat for every stage or environment we want to deploy to. The only change is the name of the job.

Now we can add the steps needed per stage.

Downloading the artifact (GH only)

In GitHub Actions, we manually have to add a step to download the artifact of the build solution workflow to be able to use it in the following steps. The YAML for this looks like the following:

```
- name: Download Artifact
  uses: actions/download-artifact@v4
  with:
    path: '${{runner.temp}}/artifacts'
    name: 'drop'
    github-token: ${{ github.token }}
    repository: ${{ github.repository }}
    run-id: ${{ github.event.workflow_run.id }}
```

Installing tooling

As was the case for the previous pipelines, we have to install the tooling.

The YAML for ADO looks like this:

```
- task: PowerPlatformToolInstaller@2
  displayName: 'Power Platform Tool Installer '
```

The YAML for GH looks like this:

```
- name: Install Power Platform Tools
  uses: microsoft/powerplatform-actions/actions-install@v1
```

Importing the solution

The next step is to import our managed solution to the target environment.

ADO

You have to alter the PowerPlatformSPN and SolutionInputFile parameters in your setup:

```
- task: PowerPlatformImportSolution@2
    inputs:
      authenticationType: 'PowerPlatformSPN'
      PowerPlatformSPN: '<Name of the ADO Service connection>'
      SolutionInputFile: '$(Pipeline.Workspace)\buildPipeline\
      drop\<Solution Schemaname>_managed.zip'
      AsyncOperation: true
      MaxAsyncWaitTime: '60'
      HoldingSolution: true
```

GH

You have to change the solution-file parameter to match the setup of your project:

```
- name: Import Solution
  uses: microsoft/powerplatform-actions/import-solution@v1
  with:
    environment-url: ${{ vars.URL }}
    app-id: ${{ vars.CLIENTID }}
    client-secret: ${{ secrets.CLIENTSECRET }}
    tenant-id: ${{ vars.TENANTID }}
    solution-file: '${{runner.temp}}/artifacts/<Solution Schemaname>
    _managed.zip'
```

```
        activate-plugins: true
        run-asynchronously: true
        import-as-holding: true
```

As you can see in the configuration, we import the solution as a `HoldingSolution`. This is needed to perform an upgrade together with the following **Apply** step. There is a new `StageAndUpgrade` configuration, which does both in one step. Depending on your setup, this might be easier and faster.

Apply Upgrade

Since we have imported the solution as a holding solution, we have to perform an Apply Upgrade step to complete the import process.

The YAML for ADO looks like this:

```
- task: PowerPlatformApplySolutionUpgrade@2
  inputs:
    authenticationType: 'PowerPlatformSPN'
    PowerPlatformSPN: ' <Name of the dev ADO Service connection>'
    SolutionName: ' <Solution Schemaname>'
    AsyncOperation: true
    MaxAsyncWaitTime: '60'
```

The YAML for GH looks like this:

```
- name: Upgrade Solution
  uses: microsoft/powerplatform-actions/upgradde-solution@v1
  with:
    environment-url: ${{ vars.URL }}
    app-id: ${{ vars.CLIENTID }}
    client-secret: ${{ secrets.CLIENTSECRET }}
    tenant-id: ${{ vars.TENANTID }}
    solution-name: ,<Solution Schemaname>'
```

In this section, we implemented the last pipeline of the source code-centric approach, the Release pipeline. With this, we are able to successfully release a solution to a downstream environment.

Running a pipeline

In this section, we will learn how to run a pipeline once it is created. Depending on how your pipeline is set up, it might start automatically (on a schedule or when some event happens), or you may have to start it manually. This section will show you how to start a pipeline that must be run manually. As always, the process is slightly different between ADO and GH.

ADO

To run a pipeline in ADO, navigate to the **Pipelines** item within your project, which will bring you to the list of recently run pipelines.

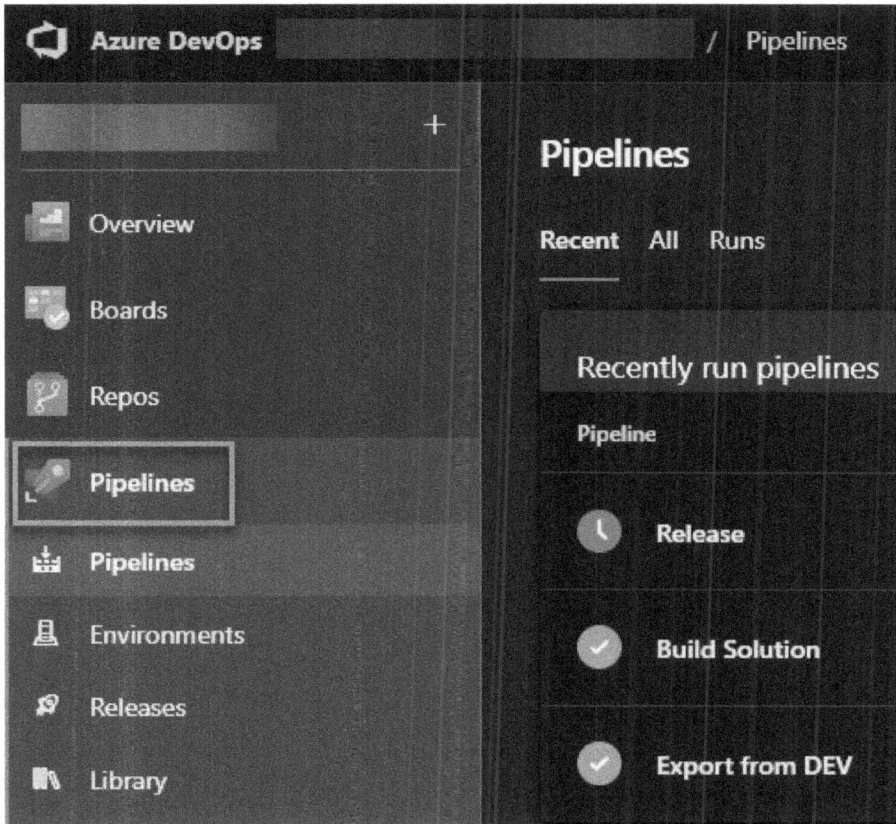

Figure 10.11: List of pipelines in ADO

From that list, you can select the pipeline you would like to run by clicking the name.

> **Not-yet-run pipelines**
> If you have a pipeline that you never have run before, you need to switch to the **All** tab to see it.

Clicking the name of a pipeline in the list brings you to a list of runs for the selected pipelines. The information of every run includes status, date and time, who triggered it, and more. Here, you also can select **Run pipeline** to trigger it manually.

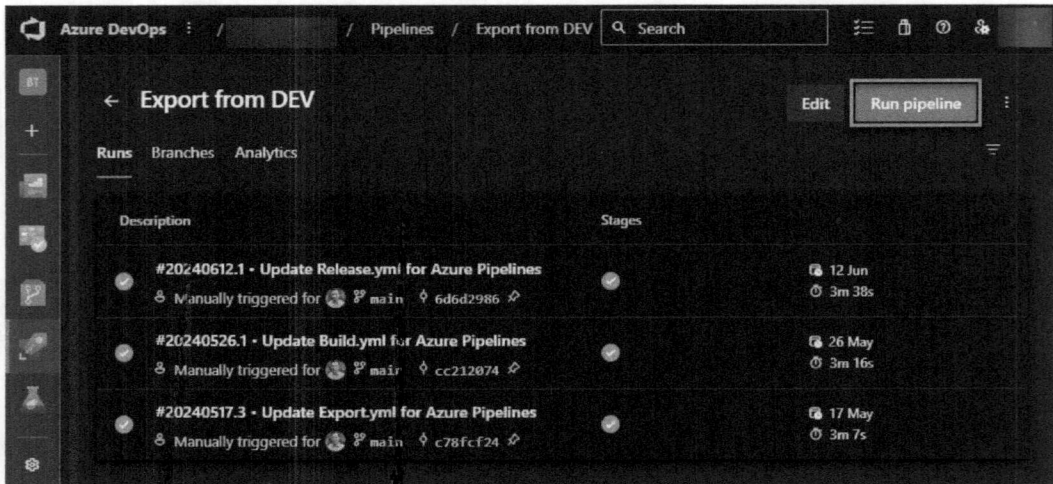

Figure 10.12: Run pipeline in ADO

Using the **Run pipeline** button will show a side pane where you can set all the parameters of the pipeline and run it.

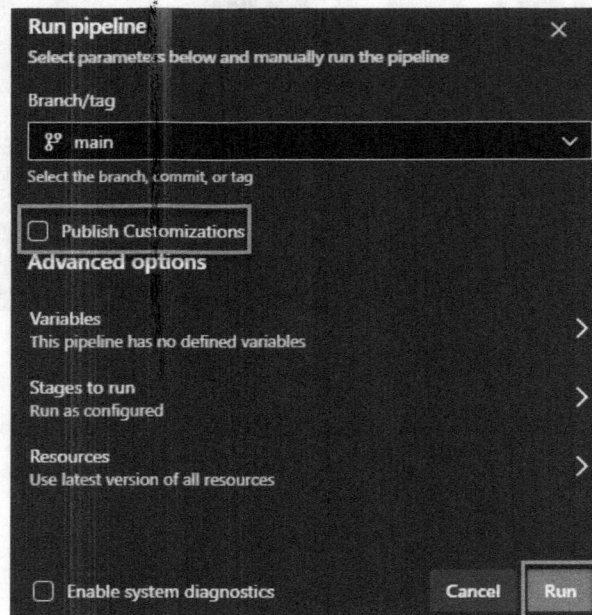

Figure 10.13: Running the pipeline in ADO

That is everything you need to know on how to start a pipeline manually in ADO.

GH

To run a pipeline manually in GH, navigate to the **Action** tab of your project. Here, you can find a list of all the workflows in your project on the left side.

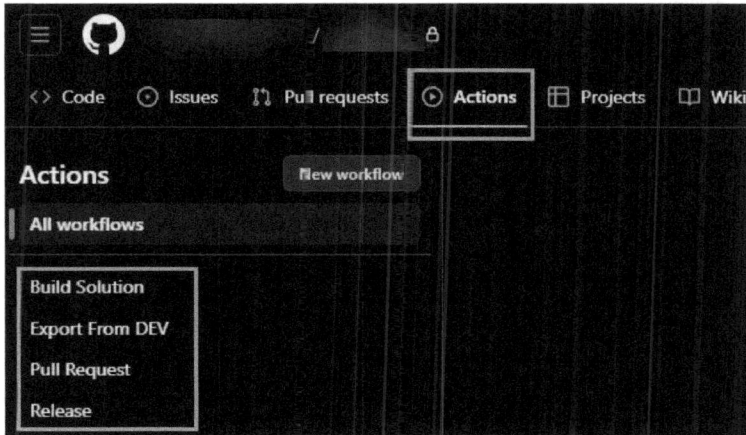

Figure 10.14: List of pipelines in GH

By selecting one of them, you will see a list of all the previous runs. The list includes detailed information on every run such as status, date and time, who triggered it, and more.

At the very top, there is a blue row where a **Run workflow** button can be found on the far-right side. Clicking the button will show a flyout where you can present input to the defined parameters, as well as run the workflow.

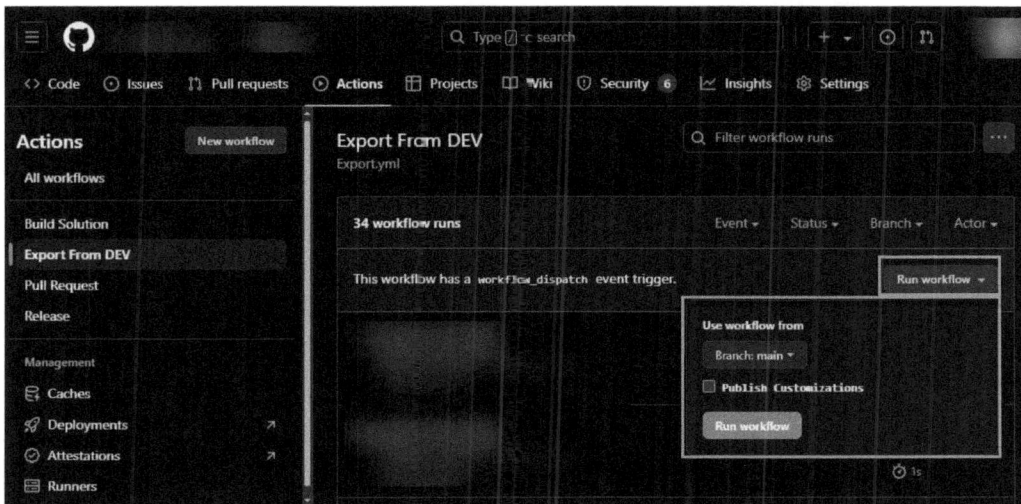

Figure 10.15: Run workflow in GH

That is everything you need to know on how to start a pipeline manually in GH.

Environment-centric approach

The second approach we discussed earlier is the environment-centric approach. PPP only supports this approach. This section will briefly describe how one can set up PPP to implement this approach.

We have already discussed the general setup of PPP. This includes how to install the solutions and how to link environments to them. Now, we will set up a simple pipeline deploying from our development environment to test.

To achieve this, we open the **Deployment Pipeline Configuration** model-driven app in our host environment. Here, we navigate to **Pipelines** and use the **+ New** button to create a new pipeline.

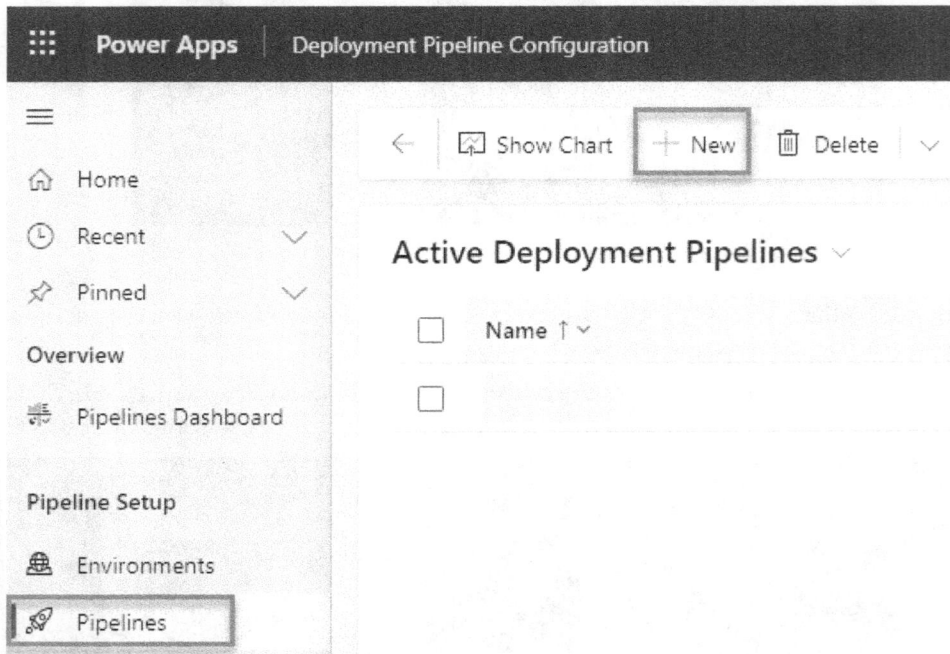

Figure 10.16: Creating a new pipeline in PPP

On the following screen, we specify a name and description. The other configurations can be as they are for the moment. Then we press the **Save** button and navigate to **Deployment stages**.

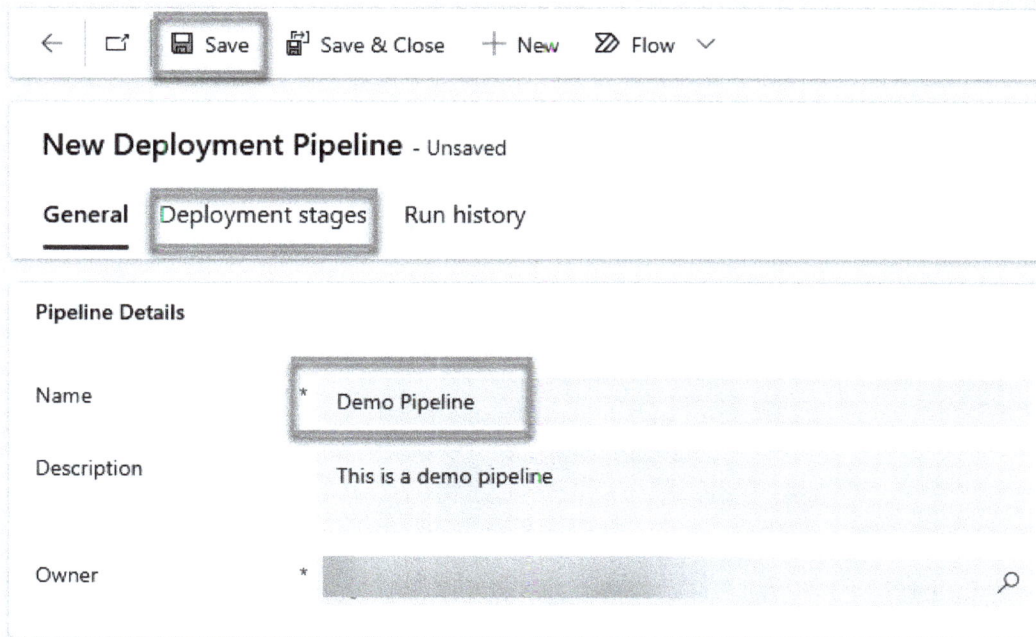

Figure 10.17: The Save button

On the **Deployment stages** tab, we can create a new stage by using the **+ New Deployment Stage** button.

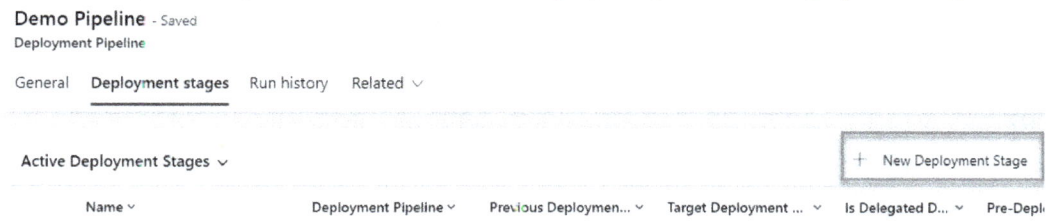

Figure 10.18: New Deployment Stage

On the following screen, we have to give the stage a name (for example, `Test`) and link it to one of the previously created target environments. We will leave all the other configurations as they are for the moment.

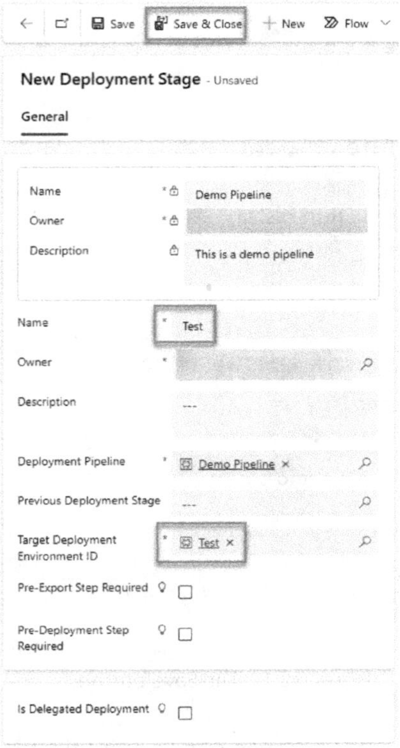

Figure 10.19: Configuring the stage

Back on the **Pipeline** record, we also have to link the development environment. Navigate to the **General** tab and select **Add Existing Development Environment** on the first subgrid.

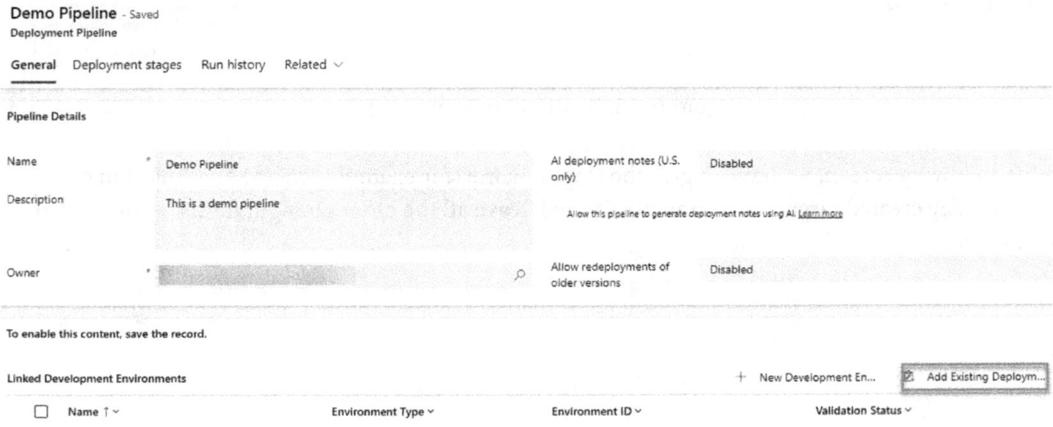

Figure 10.20: Add Existing Deployment Environment

On the upcoming side pane, we select the previously created DEV environment.

Now, we can navigate to the development environment in the maker portal. When we open a solution, the pipeline will be shown under the **Pipeline** menu.

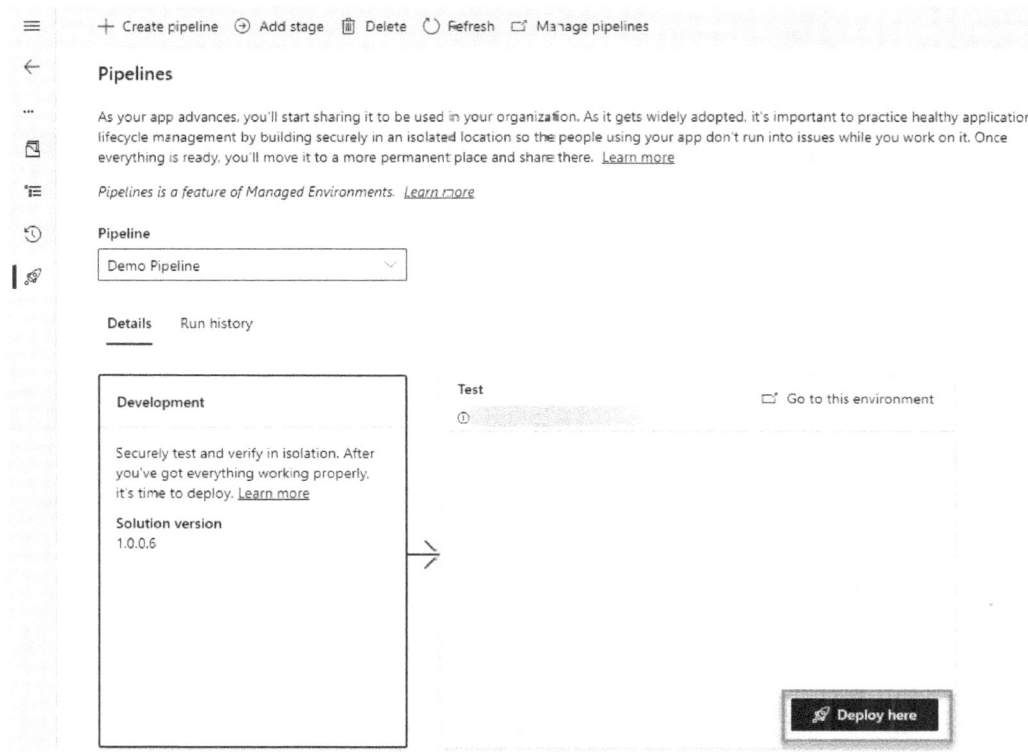

Figure 10.21: Executing the PPP

We have successfully configured the PPF to implement the environment-centric approach

Summary

As we have seen in this chapter, there are quite some steps involved in creating even a very basic ALM process. Once done, it will make deployments much more smooth, faster, and easier. We also discussed that this approach will increase the quality of your delivery by executing the correct steps, always in the same order.

In the following chapter, we will look at additional functionalities that could be added to a pipeline or workflow depending on the project's needs. This could be to handle several solutions, make the process more dynamic by giving users who run the pipeline or workflow more options, or inject code into the pack solution step.

Questions

1. Which platform can't implement a source code-centric approach?

    A. PPP

    B. GH

    C. ADO

2. What is an additional step in the release process only needed in GH?

    A. Install tooling

    B. Apply upgrade

    C. Download artifact

3. Which step is needed in the build solution process?

    A. Pack solution

    B. Export as managed

    C. Publish artifact

Further reading

-   PPP: https://learn.microsoft.com/en-us/power-platform/alm/pipelines
-   ADO YAML reference: https://learn.microsoft.com/en-us/azure/devops/pipelines/yaml-schema/?view=azure-pipelines
-   GHA workflow syntax: https://docs.github.com/en/actions/using-workflows/workflow-syntax-for-github-actions

11

Advanced Techniques

In this chapter, we will dive into some techniques to make the basic process we created in the previous chapter more resilient and usable. Not every setup will need all of them, but some are likely to be necessary.

After this chapter, you will know what settings files are, how to create and use them, how to ensure a healthy code state, as well as some methods to make YAML pipelines more robust and dynamic.

The following key topics will be covered in this chapter:

- Exploring advanced YAML
- Understanding settings files
- Healthy code state
- Transporting data

Let's get started!

Technical requirements

The following things are required to implement what we will learn in this chapter:

- An Azure DevOps or GitHub project
- At least two Dataverse environments
- A service principal added to the Dataverse environments
- Implementation of the pipelines/workflows from the previous chapter

Exploring advanced YAML

In the first part, we will take a look at advanced YAML methods. Specifically, we will discuss variables, parameters, loops, and conditions. Those are important to make your pipeline more robust and dynamic. By using them, your pipelines can handle more or less any scenario and requirement your project could encounter. You could even easily create template pipelines for reuse in different projects, without the need to redo them for each of those projects.

Variables and parameters

So far, the pipelines we created in *Chapter 10* have been very static. For example, the name of the solution was hardcoded. This means the pipeline isn't reusable without changing some parts. Another part that usually should be dynamic is whether we want to perform an update or upgrade when importing solutions, or use settings files or not.

We will learn more about settings files in the next section.

When it comes to pipelines, there are two different types of dynamic values – variables and parameters:

- *Variables* are defined and set at the beginning of the YAML code and can then be reused later.
- *Parameters* are also defined at the beginning of the YAML code but the values are set when the pipeline/action is started. They can then be reused later in the YAML code as well.

Both **GitHub (GH)** Actions and **Azure DevOps (ADO)** pipelines have the concept of variables and parameters. Both systems, use different terminology for variables and parameters, and they are also created and used differently.

As an example, we will add a variable, holding the solution name, and a parameter, defining whether the export pipeline should execute a publish customization command.

> **Note**
> Both GH Actions and ADO can handle very complex scenarios when it comes to variables and parameters. For example, GH can have variables/inputs on different levels (for a whole workflow, for a stage within a workflow, or environments). You can read more about the details in the documentation of ADO (https://learn.microsoft.com/en-us/azure/devops/pipelines/process/variables), (https://learn.microsoft.com/en-us/azure/devops/pipelines/process/runtime-parameters) and GH (https://docs.github.com/en/actions/learn-github-actions/variables).

In the following code snippets, we will add the mentioned parameters and variables to our export pipeline, as well as using the SOLUTION_NAME variable.

In the next section, we will learn how to use the publishCustomizations parameter in a condition.

We have to add the following lines to our export pipeline:

ADO

After the *pool* configuration, we will add the following lines:

```
variables:
- name: SolutionName
  value: "DemoSolution"

parameters:
  - name: publishCustomization
    displayName: Publish Customizations
    default: False
    type: boolean
    values:
      - False
      - True
```

To use the `SolutionName` variable, we will replace all the instances of our solution name with `${{variables.SolutionName}}`.

When running the pipeline, we now have the following checkbox:

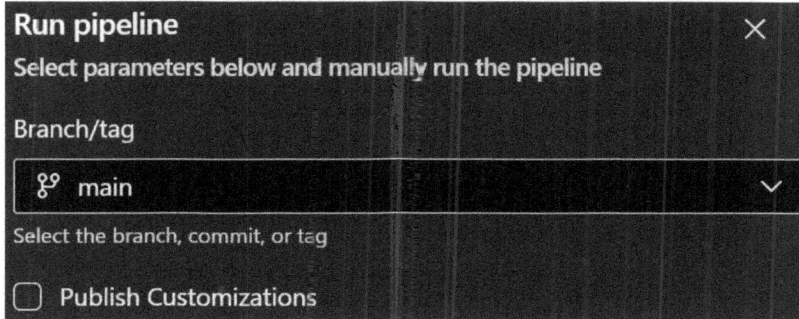

Figure 11.1: The ADC pipeline parameter

The parameter mapped to the checkbox will later be used in the pipeline to execute or skip the step of publishing customizations.

ADO has different syntax when using variables and parameters, depending on when in the execution pipeline the value should be replaced. Learn more at https://learn.microsoft.com/en-us/azure/devops/pipelines/process/variables?view=azure-devops&tabs=yaml%2Cbatch#runtime-expression-syntax.

GH

The on: part of our export pipeline from *Chapter 10* has to be replaced by the following:

```
on:
  workflow_dispatch:
    inputs:
      publishCustomizations:
        type: boolean
        description: Publish Customizations

env:
  SOLUTION_NAME: "DemoSolution"
```

To use the SOLUTION_NAME variable, we replace all the occasions of our solution name with ${{env. SOLUTION_NAME}}.

When running the workflow, we now have the following checkbox.

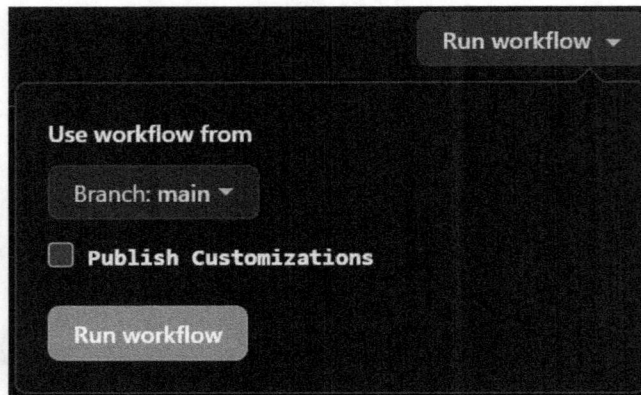

Figure 11.2: The GH action parameter

The parameter mapped to the checkbox will later be used in the pipeline to execute or skip the step of publishing customizations.

In this section, we learned how to variables and parameters both for ADO pipelines and GH Actions. Those will then be used in one of the following sections to add conditions to the YAML.

Variable file

There might be a possibility to reference a file that includes variables.

This can be interesting when you want to reuse the same variable/value in different pipelines – for example, the names of the solution(s) the pipeline should handle.

ADO

To achieve this in ADO, we have to create a new file containing the required variables:

```
variables:
- name: DemoVariable
  value: false
```

This file can then be used in the pipeline, for example, our export, build, and release pipelines, by using the `template` statement.

```
variables:
  - template: DemoVariables.yml
```

With the preceding code, we can, as mentioned, define variables that can be reused in several pipelines. This makes the handling of pipelines easier and more constant.

GH

In GH, this process is more complex, unfortunately. To achieve this, we would need to use a third-party step to read a `.env` file containing the required variables.

Now, we know how to define parameters and variables in both ADO pipelines and GH Actions. The next section will show how we can use conditions to change the behavior of automation. This is often done by evaluating variables and parameters.

Conditions

There are cases where some steps only need to run depending on certain conditions or combinations of configurations.

To achieve this, both ADO and GH have conditions. In this context, *conditions* mean that certain parts of the YAML definition only are executed when the condition is met. Otherwise, the part where the condition is placed will be skipped. We can add conditions not only to steps but also to bigger parts of the YAML definition, such as jobs.

In the following examples, we will run the publish customization step based on the parameter we created earlier. As always, the syntax between ADO and GH is slightly different.

> **Note**
> Both ADO and GH treat all parameters as strings. This is why we have to compare the value of a parameter to a `true` string instead of a Boolean.

First, we will look at an example of ADO.

ADO

For ADO, we have to add the following line to the publish customization step:

```
Condition: eq(${{parameters.publishCustomization }}, 'true')
```

When we run the pipeline without the checkbox checked, we can see that the **Publish Customizations** step is skipped.

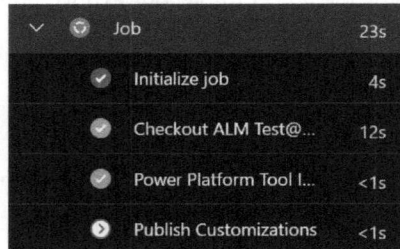

Figure 11.3: The ADO Publish Customization step skipped

Note that conditions can be on different levels – for example, on a step, stage, or job.

Next, let's look at the code example for GH.

GH

We have to add the following line to the publish customization step:

```
if: ${{ inputs.publishCustomizations == 'true' }}
```

When we now run the workflow without checking the checkbox, the publish solution step will be skipped. This could be helpful if you are sure you have done that manually before. Mostly, it is an example that is easy to understand.

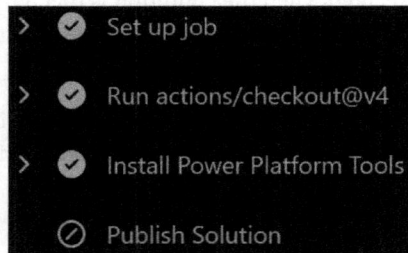

Figure 11.4: The GH Publish Solution step skipped

In this section, we have learned how we can conditionally execute steps, stages, or jobs. The next section will show how to use loops in the YAML definition.

Loops

For some scenarios, we may want to create a loop of steps to be executed several times. One example could be that a pipeline usually handles not only one solution but several at once. If so, we would want to present a list of solutions that can then be processed.

This is where *loops* come into play.

In the following example, we will add a loop to our export pipeline to generate an export of several solutions.

ADO

To implement the loop, we change the value of our `SolutionName` variable to a comma-separated list of solutions.

The next step is to add the following line before the first export step:

```
- ${{ each solution in split(variables.solutionName, ',')}}:
```

The preceding code above will perform a split on the string and then loop over the array of solution names.

For the following three steps (export as unmanaged, export as managed, and unzip), we have to increase indentation so that they are *under* the preceding `loop` statement.

We also replace all occasions of our solution name (for example, `${{variable.SolutionName}}`) with `${{ solution }}`. You can read more about expressions in YAML files on the **Microsoft Learn** page: https://learn.microsoft.com/en-us/azure/devops/pipelines/process/expressions.

GH

In GH Actions, this process is not as straightforward, unfortunately. To achieve the same, we add the following snippet to our export YAML file after the environment configuration (*line 16* in `environment: DEV`).

```
strategy:
    matrix:
        solution: ["DemoSolution", "SecondDemoSolution"]
    max-parallel: 1
```

This will run all steps within the job for every solution in the list. Since Power Platform can only handle one export at a time per environment, we have to make sure that the jobs for the different solutions run in sequence and not in parallel. This is what `max-parallel:1` is for.

We then also have to replace all occurrences of our solution (at this point, it should be `${{ env. SOLUTION_NAME }}`) with the following:

```
${{ matrix.solution }}
```

In the preceding example, the solutions are hardcoded to our pipeline. If they should be reused in different workflows, the process would be slightly different and more complex.

In addition, we have to change our SOLUTION_NAME variable to be a JSON array of the needed solutions, such as the following:

```
SOLUTION_NAME: '["DemoSolution", "SecondDemoSolution"]'
```

Then, we must add the following job as the first one in our export workflow definition:

```
variables:
    name: Compute outputs
    runs-on: windows-latest
    outputs:
      solutions: ${{ env.SOLUTION_NAME }}
    steps:
      - name: Echo
        if: false
        run: echo
```

This will make the environment variable accessible for the matrix configuration of the following export job.

> **Note**
> This extra job is needed because the matrix configuration has no way of accessing environment variables directly, due to how the context scoping is done in GH Actions. Read more at https://docs. github.com/en/actions/learn-github-actions/contexts#env-context.

In the export job, we have to add a dependency to the newly added job, by adding the following line of code after the environment configuration:

```
needs: variables
```

We also have to change the value of the solution part of the matrix configuration to the following:

```
${{fromJson(needs.variables.outputs.solutions)}}
```

In this section, we have learned how to extend pipelines with variables, parameters, conditions, and loops. Those can be used to make pipelines dynamic, more usable, and better to maintain.

Understanding settings files

Another important part of a healthy ALM is **settings files**. Those can be used to automatically set a value for environment variables, as well as the correct connection for use in connection references in the target environment. This will minimize manual work after a deployment or after a certain number of custom scripts have been used.

In this section, you will learn how to generate settings files and how to use them in a pipeline.

> **Note**
>
> Microsoft is investing in this area as well. There are plans to introduce more settings to the settings files. For example, there might be the possibility to automatically share apps with the correct groups of people through configuration in these files in the future.

Generating a settings file

First of all, the settings file has to be generated.

To do this, the `pac` CLI (as described in *Chapter 5*) has to be installed. We also need an unmanaged version of our solution file on our machine.

Open a command line or PowerShell in the folder where the solution file is located.

The following command will create a setting file in the JSON format:

```
pac solution create-settings -z ./DemoSolution_1_0_0_6.zip -s ./
settings.json
```

The file will contain two arrays, `EnvironmentVariables` and `ConnectionReferences`.

For every environment, we need one file with the correct values.

The next step is to fill the file with the correct values for the environment in question.

The values for the environment variables are easy to fill, since those are mostly strings, numbers, or Booleans. For environment variables of type `secret`, we have to compose the correct URL for the secret in the key vault.

The values for the connection references are the IDs of the connections that should be used in the target environment. To get them, we have to open the details page of the connection in question and copy the ID from the URL.

You can read more about this on the Microsoft Learn page: `https://learn.microsoft.com/en-us/power-platform/alm/conn-ref-env-variables-build-tools#step-2-get-the-connection-reference-and-environment-variable-information`.

All the files (for every solution and every environment) need to be stored in the repository.

In our case (according to the folder structure discussed previously in *Chapter 9* in the *Folder structure* section), we will store them in the `PowerPlatform/Solutions/Settings` folder of our repo. The naming convention for this example is `<SolutionName>-<Environment>.json` – for example, `DemoSolution-Test.json`. You can alter the names to your needs.

> **Note**
> Some people prefer to not store the settings file(s) in the repository and, rather, create them as artifacts of the build solution. In my opinion, this makes the process much more complex and isn't necessary for most of the scenarios, even though it also is a valid approach.

Using Settings files

Now that we know how to create a settings file, we have to take a look at how to use them.

This depends on the repo structure you have for your project. In this example, we will assume the files are stored as described in the previous section.

All we have to do is add two more parameters to the import step(s) within our release pipeline.

`SolutionName` can be a variable, as described earlier in this chapter.

ADO

The following code will configure the settings file in the ADO pipeline. Make sure to replace `SolutionName` and `Environment` with the information matching your project:

```
UseDeploymentSettingsFile: true
DeploymentSettingsFile: '$(Agent.BuildDirectory)/s/PowerPlatform/
Solutions/Settings/<SolutionName>-<Environment>.json'
```

GH

The following code will configure the settings file in GH Actions. Make sure to replace `SolutionName` and `Environment` with the information matching your project:

```
use-deployment-settings-file: true
deployment-settings-file: 'PowerPlatform/Solutions/
Settings//<SolutionName>-<Environment>.json'
```

In this section, you have learned how to generate and use settings files. The next section will explain how to make sure that the code we deliver is in a healthy state.

A healthy code state

Ensuring a healthy code state is important when it comes to any development. By healthy code state, we mean that the code components (plugins, custom components (PCFs), and web resources) that get deployed to the downstream environments should be in a ready state.

Usually during the implementation, a developer deploys versions to the development environment that aren't ready to be released to the downstream environment yet.

In addition, there are two reasons why we would like to achieve a healthy code state:

- The developer could forget to deploy the latest working version to development after the implementation is completed
- Solutions to contain the binaries (.dll files) should not be checked into the repository

In this section, we will take a look at one approach to how we can make sure that we only deploy the correct version of pro code components.

There are at least two different approaches:

- **Map files**: We will take a deeper look at this approach
- Deploying the latest version to development before executing the export

Let's review the first approach.

Creating a map file

A map file will relate a file (Plugin Assembly DLL or web resource) from the solution to the corresponding file in the repository. When presented to the `unpack` command, the unpack will skip the files specified in the map file from being unpacked. If the same file is presented to the `pack` command, it will take the corresponding files from the repository and pack them into the solution where they should be, according to the configuration in the map file.

The approach of a map file can be used to skip storing binary files in your repository. as well as inject a certain version of those files into the generated solution.

First of all, we have to create our map file. There are two mapping approaches one can configure within the map file:

- **Folder mapping**: The matching folder will be replaced with the configured "to" folder
- **File-to-file mapping**: The matching file will be replaced with the configured "to" file

For our example, we will focus on file-to-file mapping and only map one Plugin Assembly.

The used path will be dependent on the structure of your project.

Let's assume we have one Plugin Assembly, `DemoPlugin.dll`. With our current unpack setup in the export pipeline, the `.dll` file will be in the following folder: `PowerPlatform\Solutions\DemoSolution\PluginAssemblies\DemoPlugin.dll`

As mentioned, the beginning of the paths depends on your project structure as well as whether you use ADO or GH.

The following is an example XML file that works for the demo set up in ADO:

```
<?xml version="1.0" encoding="utf-8"?>
<Mapping>
    <FileToFile map="D:\a\1\s\PowerPlatform\Solutions\DemoSolution\
    BebePluginsAccount-801E1333-DAA6-4FEB-BD67-AA51E1CCF4DD\
    BebePluginsAccount.dll" to="D:\a\1\s\development\Back-end\Plugins\
    Account\bin\release\BebePluginsAccount.dll" />
</Mapping>
```

This file then needs to be placed in the repository.

In our case (according to the folder structure discussed previously), we will store them in the `PowerPlatform/Solutions/Mappings` folder of our repo. The naming convention for this example is `<SolutionName>.xml` – for example, `DemoSolution.xml`. You can alter the names to your needs.

Read more about mapping files on the Microsoft Learn page: `https://learn.microsoft.com/en-us/dynamics365/customerengagement/on-premises/developer/compress-extract-solution-file-solutionpackager` `#use-the-map-command-argument`.

Using a mapping file

The mapping file has to be used both in the unpack step in the export pipeline, as well as the pack step in the build solution pipeline.

The syntax is the same for both steps, but it is slightly different in ADO and GH.

We only must add one line, which specifies the path that the map file should use. The unpack/pack steps will do everything else themselves.

> The only thing we have to make sure of when it comes to the pack step is that we build the plugin project before the pack step is executed. This is to make sure that the `.dll` file is also present on the pipeline/workflow agent/runner.

The following code snippets show how to implement this approach in ADO and GH. Make sure to replace `<SolutionName>` with the value you need for your project.

ADO

The following line is the one we have to add to an ADO pipeline:

```
MapFile: ' $(build.sourcesdirectory)\PowerPlatform\Solutions\
Mappings\<SolutionName>.xml'
```

GH

The following line is the one we have to add to a GH workflow:

```
map-file: 'PowerPlatform/Solutions/Mappings/<SolutionName>.xml'
```

Transport data

As we described in *Chapter 6*, it is sometimes necessary to transport configuration or reference data between environments. This section will show you how that can be done in a pipeline.

We will use the **Data Migration Utility** (**DMU**) from Microsoft to do that. This is, as described in *Chapter 6*, the preferred way, since Power Platform Build Tools already have steps to execute DMU schemas.

Creating a schema

First of all, we have to create a schema file. To do this, we open the DMU and connect to our dev environment. At the top, we can select a solution (in our case, **Demo Solution**). After that, the list of available tables will be filtered to those within the selected solution.

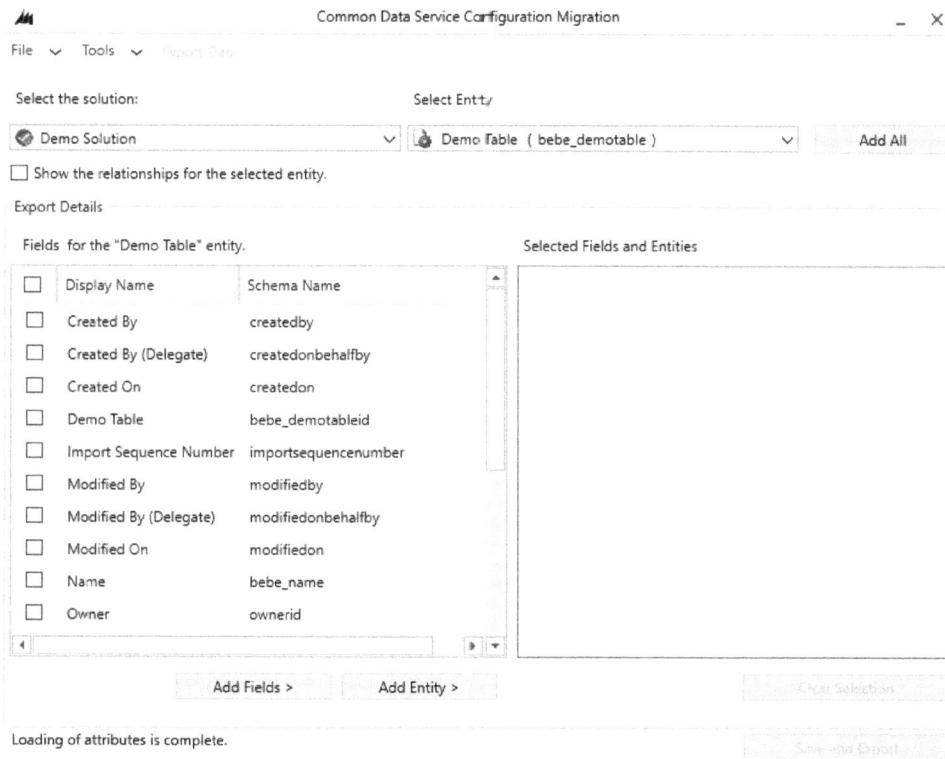

Figure 11.5: Creating a DMU schema – selecting a solution

Then, you add the tables you want to export as well as the fields of those you want to export. As an example, you can see in the following screenshot that we have added **Demo Table** with all its fields.

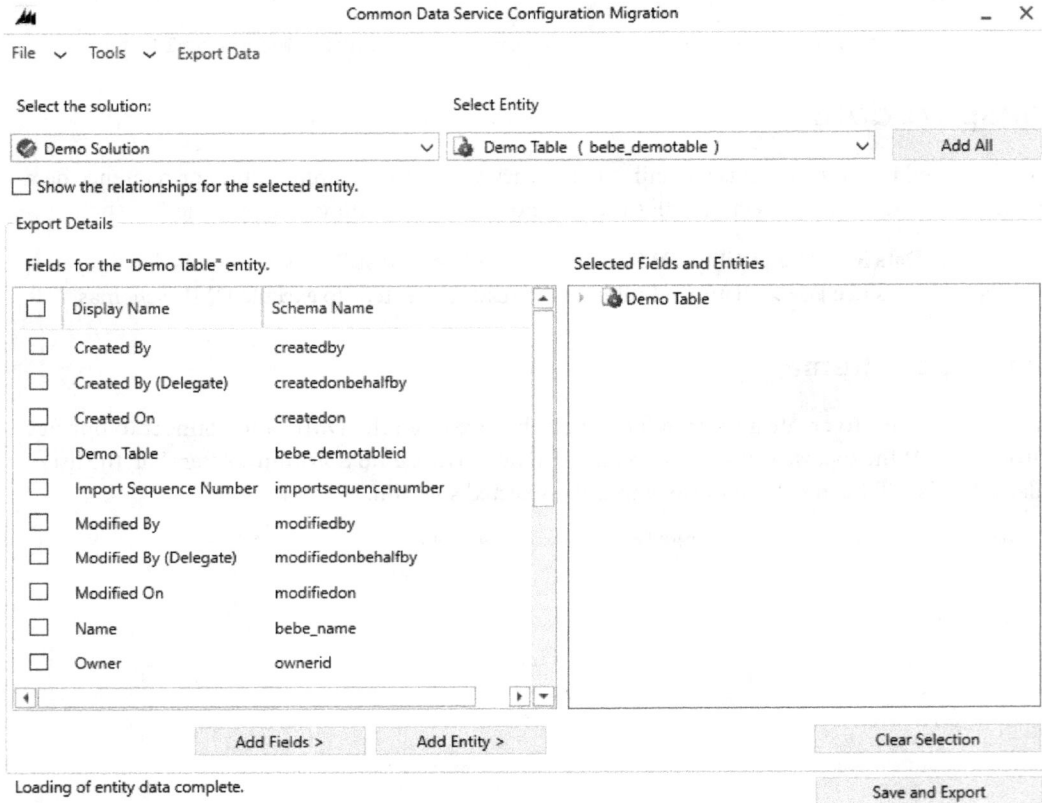

Figure 11.6: Creating a DMU schema – selecting a table

After clicking on **Save and Export**, you have the chance to save the created schema as an XML file, as well as do an initial export to see whether everything works as expected.

We save the schema file as `DemoSolution.xml` in the `PowerPlatform\Data\Schema` folder in our repo.

This file is later used to export the configured tables as well as import downstream environments.

Using a schema

Now that we have created the schema in the previous section, we have to use it.

In this example, we will create an export in the build pipeline and store it as part of our pipeline artifact. In the release pipeline, we will then use this artifact and import it to the downstream environments. The first part, generating the export, could be skipped if you generate the export manually and store the ZIP file in the repository as well.

Generating an export of data

To generate the export of data, we will add the following step between the pack and the publish artifact steps in our build pipeline.

ADO

The following snippet will, as mentioned, add a step to export the data we configured in the schema:

```
- task: PowerPlatformExportData@2
  inputs:
    authenticationType: 'PowerPlatformSPN'
    PowerPlatformSPN: 'DEV'
    Environment: '$(BuildTools.EnvironmentUrl)'
    SchemaFile: '$(build.sourcesdirectory)\PowerPlatform\Data\Schema\
    DemoSolution.xml'
    DataFile: '$(Build.ArtifactStagingDirectory)\\DemoSolution-data.
    zip'
    Overwrite: true
```

GH

The following snippet will, as mentioned, add a step to export the data we configured in the schema:

```
- name: Export Data
  uses: microsoft/powerplatform-actions/export-data@v1
  with:
    environment-url: ${{ vars.URL }}
    app-id: ${{ vars.CLIENTID }}
    client-secret: ${{ secrets.CLIENTSECRET }}
    tenant-id: ${{ vars.TENANTID }}
    schema-file: 'PowerPlatform\Data\Schema\DemoSolution.xml'
    data-file: '${{runner.temp}}/artifacts/DemoSolution-data.zip'
    overwrite: true
```

Now that we have our exported data as an artifact, we can change the release pipeline to also use it.

Importing data

Next, we have to add the following step to our release pipeline to import the data we exported earlier. This should be done as the last step in the pipeline.

ADO

The next snippet adds the import step to an ADO pipeline:

```
- task: PowerPlatformImportData@2
  inputs:
    authenticationType: 'PowerPlatformSPN'
    PowerPlatformSPN: 'Test'
    Environment: '$(BuildTools.EnvironmentUrl)'
    DataFile: '$(Pipeline.Workspace)/buildPipeline/drop/DemoSolution-
    data.zip'
```

GH

The next snippet adds the import step to a GH pipeline:

```
- name: Import Data
  uses: microsoft/powerplatform-actions/import-data@v1
  with:
    environment-url: ${{ vars.URL }}
    app-id: ${{ vars.CLIENTID }}
    client-secret: ${{ secrets.CLIENTSECRET }}
    tenant-id: ${{ vars.TENANTID }}
    data-file: '${{runner.temp}}/artifacts/DemoSolution-data.zip'
```

Summary

In this chapter, we have learned how we can extend a basic pipeline to the needs of a specific project. All of the shown techniques can be, and most likely will be, combined. This can create a very robust and dynamic automation process that can handle more or less every scenario.

In the next chapter, we will learn the process of completing continuous integration/continuous delivery, which includes quality gates, automated testing, and versioning.

Questions

1. Which approach is used to make a pipeline more dynamic?

    A. Variables/parameters

    B. Loops

    C. Mapping a file

2. What is used to change the pipeline/workflow behavior while starting it?

    A. Variable

    B. Parameter

    C. Condition

3. Which component makes it possible to reuse information in several pipelines/workflows?

    A. A parameter

    B. A settings file

    C. A variable

Further reading

- Conditions: https://learn.microsoft.com/en-us/azure/devops/pipelines/process/conditions

- Using conditions: https://docs.github.com/en/actions/using-jobs/using-conditions-to-control-job-execution

- Expressions: https://learn.microsoft.com/en-us/azure/devops/pipelines/process/expressions

- Using jobs: https://docs.github.com/en/actions/using-jobs/using-a-matrix-for-your-jobs

- Connections and Connection References explained: https://benediktbergmann.eu/2022/02/08/connections-and-connection-references-explained/

- Working with Connection References: https://benediktbergmann.eu/how-to-use-connection-references

12
Continuous Integration/ Continuous Delivery

When it comes to classic software development, the intention is to create a system where **continuous integration (CI)** and **continuous delivery (CD)** are possible.

Continuous integration aims to automatically and frequently integrate code changes into a repository. Continuous delivery is about delivering code changes to production in a frequent manner.

This chapter will cover several topics that, in combination, make it more feasible to introduce a CI/ CD process for a Power Platform project.

We'll cover the following topics:

- Branching
- Quality gates
- Versioning

We will learn about branching and different quality gates for pipelines, as well as discuss versioning when it comes to the Power Platform.

Using those techniques can improve the overall quality of a delivery. In addition, the versioning increases clarity in regard to which version of the implementation is deployed to which downstream environment and is needed to make use of the platform capability to perform a solution upgrade.

Branching

Branching is a very common concept when it comes to the development of any software. In the realms of the Power Platform, it is not used very often though.

As briefly mentioned in *Chapter 7*, the source-code-centric approach makes it possible to work with branches when it comes to development for the Power Platform.

The general idea is to have at least one branch that always has a working version of the software in question. In addition, there can be other different branches that contain varying versions. Those versions might or might not be working. This is to support the several development phases software might have.

Types of branches

There are several approaches to branching for classic development – for example, using Git Flow, a Trunk-based, or an environment-based branching strategy.

Git Flow basically has two branches. One is called main, which contains the released versions of software, and one is called Develop, which contains the current state of the development. In addition, there could be feature branches that contain unfinished work. The Develop branch also serves as an integration branch for the mentioned features. In this approach, branches are long-lived.

A trunk-based strategy uses short-lived branches and very frequent commits. The development trunk or branch is always considered stable.

For the Power Platform, I found that some versions of environment-based branching suit most projects that need branching the best. It is a good balance between keeping the process clean and easy, as well as offering enough possibilities and flexibility. The following section describes how environment-based branching works and how it can be used on the Power Platform.

A well-designed branching approach is an important step in coming closer to a CI/CD setup because it is important to separate working code and customization from non-working code and customization already in the repository.

Environment-based branching

As the name suggests, this approach aims to have one branch per environment. Since we work with environments on the Power Platform a lot, this approach suits most Power Platform projects.

You basically have one branch per environment where the master/main branch is related to your production environment.

From the main branch, we would create a test branch, and from there a develop branch.

Whenever we would like to deploy from DEV to Test, we could merge it back to the Test branch. This would trigger a build pipeline to create a new version of the solution and deploy it to the Test branch. The same would happen when we would like to deploy to production.

Quality gates

This section is about quality gates in the pipelines used in a project. With quality gates, we mean steps throughout your pipeline that stop the process moving forward if it fails. This is used to, as the name suggests, ensure a certain quality throughout the implementation.

> **Note**
>
> Even if you are not interested in or planning on implementing CI/CD in your project, I would recommend implementing quality gates. This will, as mentioned, increase the overall quality of your project.

In this section, we will look at three examples of quality gates:

- Solution checker
- Automated unit testing
- Pull requests

Depending on your project setup and requirements, there could be other quality gates.

Those could be static code analysis or automated UI testing, as two examples. But there are many more that might be of interest.

Solution checker

One of the low-hanging fruits that I recommend for every project is the Power Platform solution checker. The **Solution Checker** is a piece of software created by Microsoft. It is part of the Power Platform and is constantly updated. When it is run, it checks a solution's ZIP file against a set of Microsoft-defined rules. Those rules span web resources, plugins, flows, and Power Fx expressions.

> **Note**
>
> Microsoft has published an npm package containing those rules. This can be used in your JavaScript or TypeScript setup to run the rules locally. Using this, errors are detected much earlier: `https://powerapps.microsoft.com/en-us/blog/announcing-public-preview-of-eslint-rules-for-power-apps-and-dynamics-365/`

The idea is to take an unmanaged solution ZIP file and run the solution checker against it. This will give you a list of things, which Microsoft would consider as warnings and errors.

A solution can be checked against the solution checker directly in the maker portal or in a pipeline.

When it comes to pipelines, one could define a threshold to fail the task when exceeded. This means that the whole pipeline run will stop when the solution checker task fails.

This makes sure that our solution is always compliant with Microsoft rules and, therefore, increases quality.

To run the solution checker in a pipeline, we only have to add one step after we have exported the solution as unmanaged.

Azure DevOps (ADO)

The following YAML snippet shows how to add the solution checker to your export pipeline in ADO:

```
- task: PowerPlatformChecker@2
  inputs:
    authenticationType: 'PowerPlatformSPN'
    PowerPlatformSPN: 'DEV'
    FilesToAnalyze: '$(Build.ArtifactStagingDirectory)\\${{ solution
    }}.zip'
    RuleSet: '0ad12346-e108-40b8-a956-9a8f95ea18c9'
```

GitHub (GH)

The following YAML snippet shows how to add the solution checker to your export pipeline in GH:

```
- name: Check Solution
        uses: microsoft/powerplatform-actions/check-solution@v1
        with:
          environment-url: ${{ vars.URL }}
          app-id: ${{ vars.CLIENTID }}
          client-secret: ${{ secrets.CLIENTSECRET }}
          tenant-id: ${{ vars.TENANTID }}
          path: "${{runner.temp}}/exported/${{ matrix.solution }}.zip"
```

Automated unit testing

Another quality gate I would recommend to everyone is to write tests and run them automatically in pipelines.

> **What are tests?**
>
> An automated test is some form of code that can be run to test whether the implementation is working as expected. In general, a test consists of three parts: Prepare, Act/Execute, and Assert. First, every test has to prepare everything so that the fake database looks as expected. Then, the actual implementation of the test will be executed. Lastly, it should check whether the implementation did what was expected.

The kind of tests are very much dependent on the project. Unit tests and integration tests are suitable for most projects. In addition, there could be UI tests or performance tests, for example.

Where those automated tests are run also depends on the project setup. Usually, they are run before the current code state is used in the process of packaging the solution file. This might be in the build pipeline we described earlier or the export pipeline of an environment-centric approach before the latest version is deployed to DEV.

The basic process is to build the solution and run VSTest to run the tests. In addition to that, we also would need to do a NuGet restore (for C# projects) and npm install (for TypeScript projects).

The following code snippets show a setup that would run all tests written for any .NET Framework project in your repository. Usually, one would like to filter the tests to certain projects or specific test types. Also, the locations depend on the setup and structure of your project.

ADO

The following YAML will run all C# Unit tests in an ADO pipeline:

```
- task: NuGetToolInstaller@1
- task: NuGetCommand@2
  inputs:
    command: 'restore'
    restoreSolution: '**/*.sln'
    feedsToUse: 'select'
    vstsFeed: '6283727e-a3fa-4c12-a674-f8f805edc937'
- task: MSBuild@1
  inputs:
    solution: '**/*.sln'
- task: VisualStudioTestPlatformInstaller@1
  inputs:
    packageFeedSelector: 'nugetOrg'
    versionSelector: 'latestPreRelease'
- task: VSTest@3
  inputs:
    testSelector: 'testAssemblies'
    testAssemblyVer2: |
      **\*test*.dll
      !**\*TestAdapter.dll
      !**\obj\**
    searchFolder: '$(System.DefaultWorkingDirectory)'
```

GH

The following YAML will run all C# unit tests in a GH workflow:

```
- name: Setup MSBuild
      uses: microsoft/setup-msbuild@v1

   - name: Setup NuGet
      uses: NuGet/setup-nuget@v1.0.5

   - name: Restore Packages
      run: nuget restore MySolution.sln

   - name: Build solution
      run: msbuild MySolution.sln -t:rebuild
-property:Configuration=Release

   - name: Run tests
      run: microsoft/vstest-action@v1.0.0
```

> **Note**
>
> It might be that the latest Windows runner does not support older versions of the .NET Framework used for plugins. In that case, you must switch your pipeline to a "windows-2019" runner instead.

Pull requests

In a real-life scenario, a proper CI/CD setup should not commit changes to shared branches (like Main or develop) directly. Instead, **Pull Requests** (**PRs**) should be used. With that, a developer would create a copy of the current state in a new branch, make all the necessary changes there, and create a pull request to merge everything back to the branch in question.

The user should only be able to merge the pull request into the source branch when at least one other person has reviewed the PR, no comments are open, and all tests were run successfully. Tests should automatically be run when the PR is created and whenever a change/commit to it is made. All of this is to increase the quality of code that comes into your active branches.

To achieve this, we create a new pipeline/action that only runs our tests, as described.

Then we must add a new branch policy (ADO) or branch protection rule (GH).

ADO

To add a branch policy, we navigate to **Branches** and choose **Branch policies** from the vertical ellipsis of the master branch.

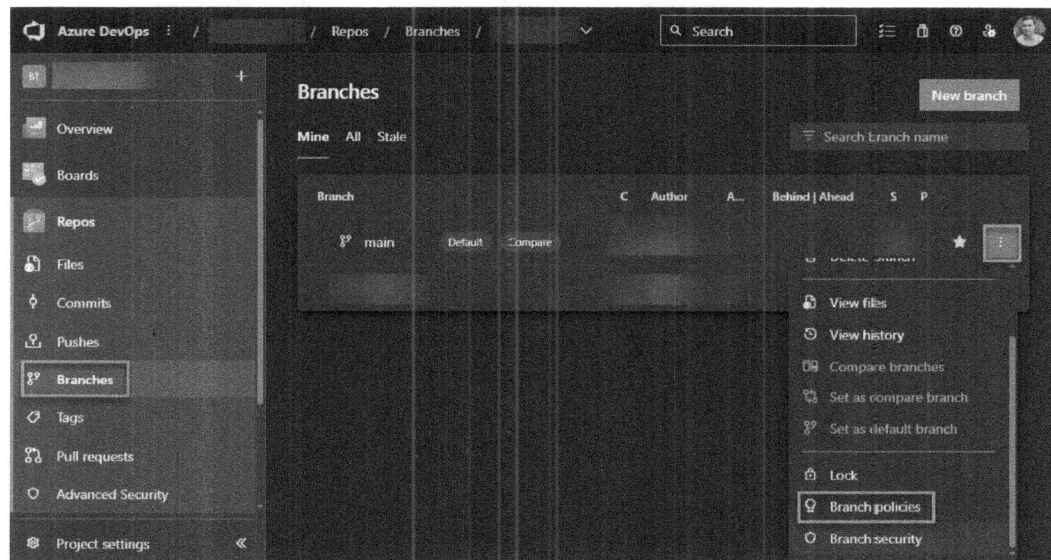

Figure 12.1 – Azure DevOps UI to access Branch policies

Here, we will check the **Require a minimum number of reviewers** box, change the minimum number to the number you would like to have (for the demo we use one), and make sure to not check **Allow requestors to approve their own changes**.

> **Note**
>
> If you check the **Allow requestors to approve their own changes** checkbox and set the minimum number of reviewers to 1, it will allow one person to merge the pull request without a check from someone else.

The second thing we check is **Check for comment resolution**. This will require all comments to be resolved (if **Required** is chosen).

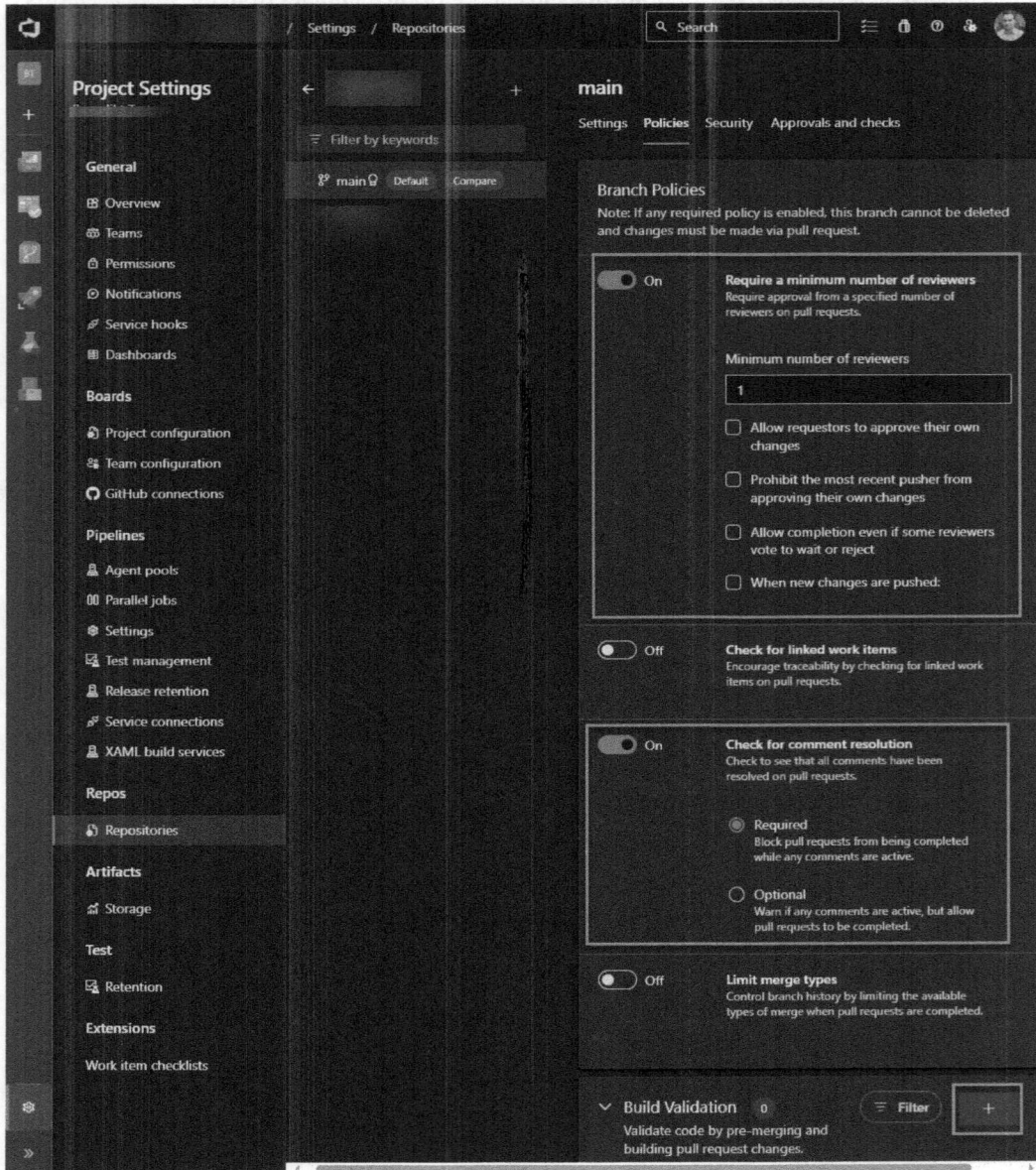

Figure 12.2 – Configure policy

The last step in configuring the policy is to add a build validation by clicking the + button (see the preceding screenshot). This will open a flyout on the right side.

Under **Build pipeline**, we choose the pipeline we created earlier. All the other configurations can be left as they are.

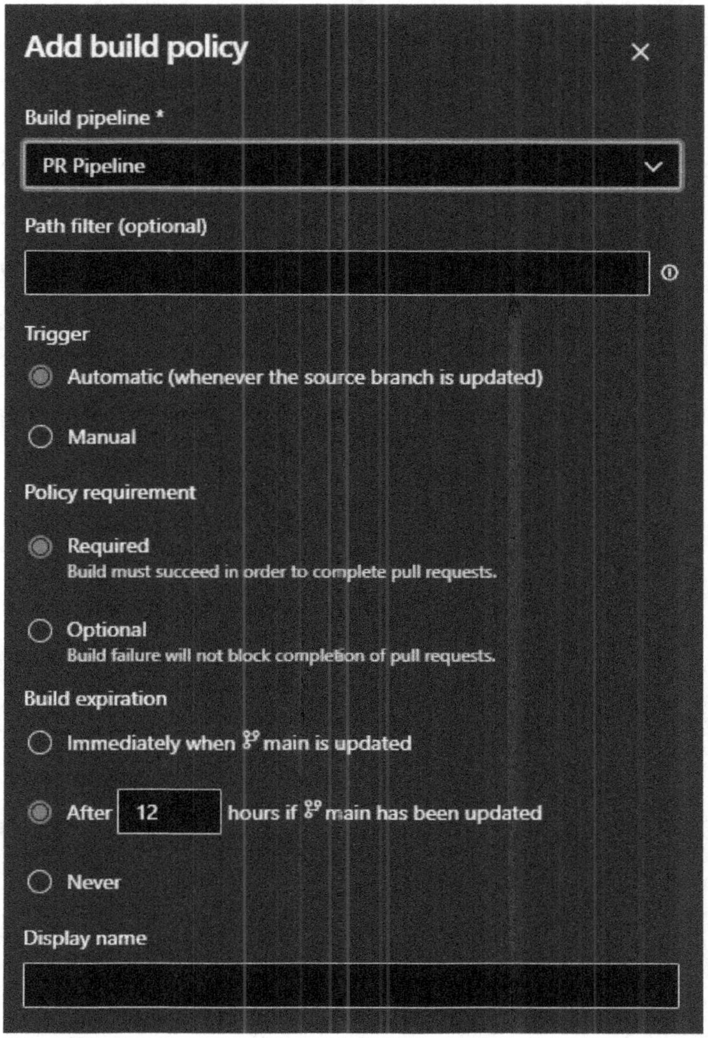

Figure 12.3 – Configure build validation

The complete policy will look like this:

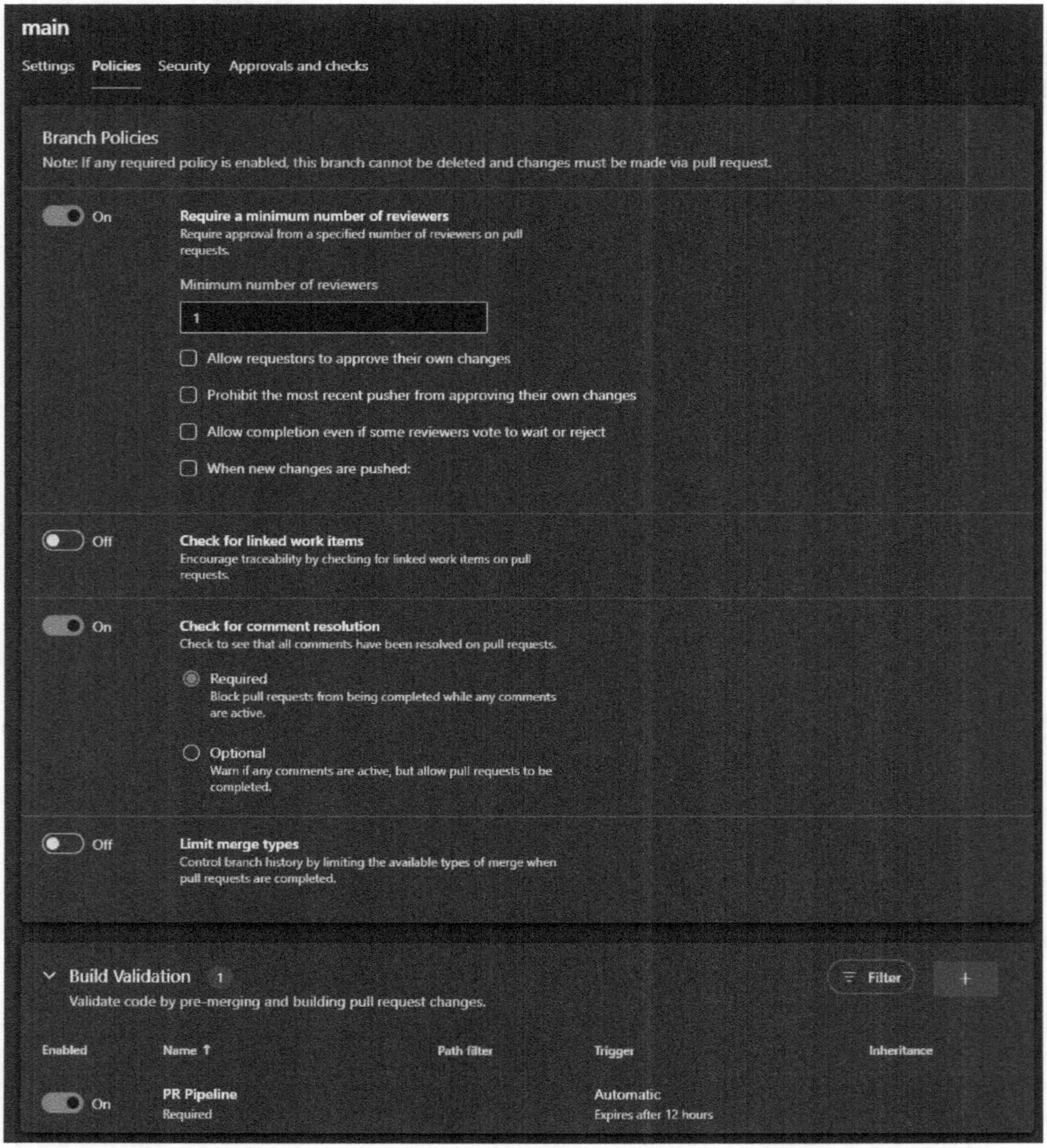

Figure 12.4 – Complete branch policy

After saving the policy, by using the **Save changes** button at the top, it is active immediately.

GH

To add a branch rule, we open our repository in GitHub and navigate to **Settings | Branches**. Then, we use the **Add branch ruleset** button.

Here, we specify a ruleset name, enable it, and add the default branch as a target:

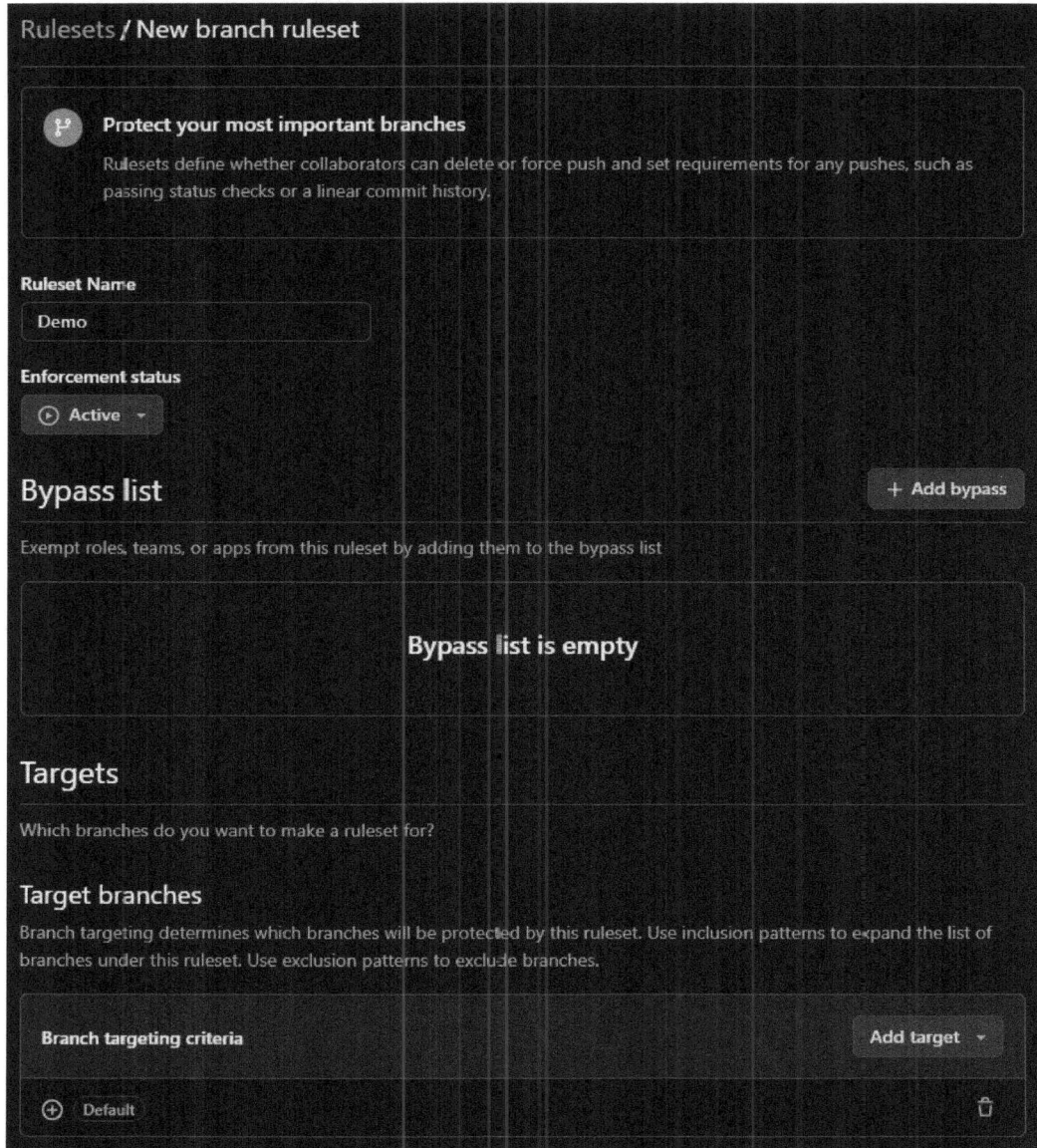

Figure 12.5 – Configure branch ruleset in GH

Under Rules, we activate **Require a pull request before merging**. In the sub-configuration, we change **Required approvals** to **1** for this example (this will vary from project to project). **Dismiss stale pull request [...]** should also be selected.

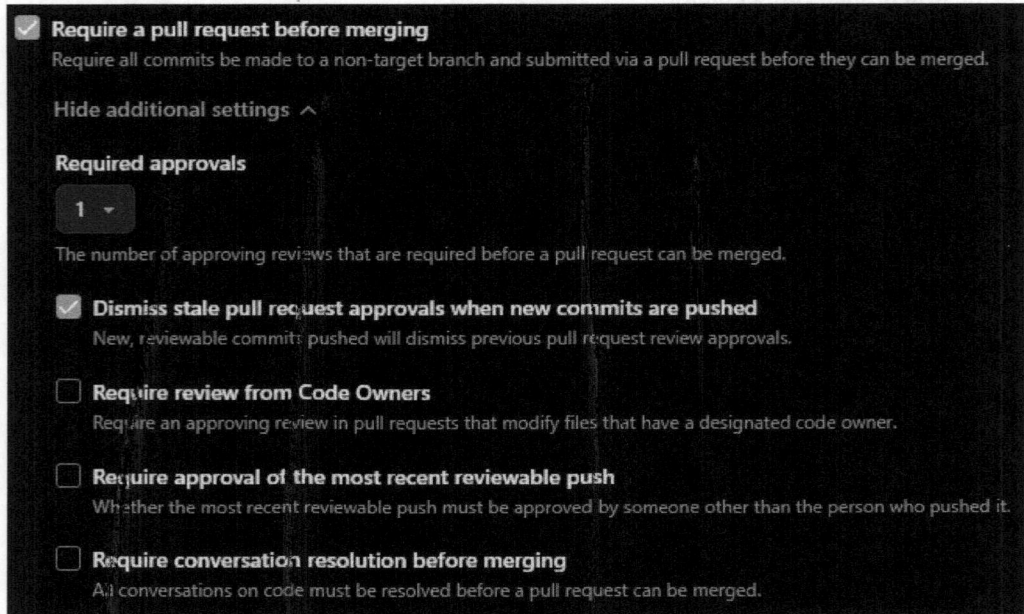

Require a pull request before merging
Require all commits be made to a non-target branch and submitted via a pull request before they can be merged.

Hide additional settings ∧

Required approvals

1 ▾

The number of approving reviews that are required before a pull request can be merged.

Dismiss stale pull request approvals when new commits are pushed
New, reviewable commits pushed will dismiss previous pull request review approvals.

Require review from Code Owners
Require an approving review in pull requests that modify files that have a designated code owner.

Require approval of the most recent reviewable push
Whether the most recent reviewable push must be approved by someone other than the person who pushed it.

Require conversation resolution before merging
All conversations on code must be resolved before a pull request can be merged.

Figure 12.6 – Adding approvals

We also activate **Require status checks to pass**. Here, we add our test pipeline as a check.

The last step is to click the **Create** button.

We also have to change the trigger of the pipeline to the following:

```
on:
  pull_request:
    branches: [ main ]
```

`main` has to be replaced with the name of the branch you'd like to secure.

In this section, we have learned how to secure our branches. From now on, we need to create pull requests and can't commit to the secured branch directly. A pull request has to be approved by at least one other developer to implement a need for cross-checks. In addition, a pipeline that executes tests is run whenever someone creates a pull request or makes changes to one. Only if the configured pipelines run through successfully is a merge of the PR possible.

Versioning

Versioning is another important topic in general, but especially when it comes to CI/CD. Versions of Power Platform solutions have four different parts. Those parts are named as follows:

- Major
- Minor
- Build
- Revision

The goal is to not only change the version of the solution but also the containing components (for example, the plugin assembly).

> **Note**
> Be aware that when the major or minor part (the first two parts) of a plugin version is changed, a new parallel version of the assembly will be created instead of replacing the older one. This is done for migration purposes but can lead to major issues if not handled.

Every project needs to decide on an approach to how to handle versioning. In general, there are the following approaches:

- Manually increase the version when needed
- Increase the version with every pipeline/action run

Of course, there are a lot of approaches in between where, for example, the version is automatically increased once a day.

The approach I have often used is that every pipeline run increases the revision part of the solution version before doing an export. The other three parts are handled through a config file. Whenever one of the first three parts changes, the revision number starts over from 0.

This approach makes it very clear which version is exported and used, but it is harder to handle since any version change in the maker portal will be overridden by the configuration we have in the config file. It could also lead to high revision numbers when the export pipeline runs frequently. In a perfect world, the version shouldn't change just because an export is run. It should only change when a new version is built to bring it to the next environment.

More recently, Microsoft introduced a flag on the import step that can skip an import if the version to import is the same or lower than the existing version of the solution in the environment. With the above-mentioned approach, this flag becomes unusable since the version is always increased automatically even when there hasn't been any change in the solution. All those reasons could lead to the use of a manual approach.

In a source code-centric approach, the recommendation is to change the version when the build pipeline is creating the new ZIP file, which will be deployed to downstream environments.

As always, this highly depends on your setup and your requirements. It is important to think about it and set guidelines for it though.

Summary

In this chapter, we have learned a lot about how to increase quality throughout the development process. This included quality gates such as Solution Checker or automated testing, as well as pull requests.

Another topic we learned about was branching, the different strategies there are, and which is used the most in Power Platform projects.

We also briefly discussed which different versioning approaches there are and some of the pros and cons they bring.

All that we learned can be used to increase the quality of the delivery.

The next chapter will give you a summary of what we have learned in this book. We will also discuss some additional best practices.

Questions

1. What do you have to add to your branch if you would like to protect your PR in ADO?

    A. Branch protection rule

    B. Branch policy

    C. Branch security

2. What is the most used branching strategy when it comes to Power Platform projects?

    A. GitFlow

    B. Trunk-based

    C. A version of GitHub Flow

3. What is a quality gate?

    A. Solution Checker

    B. Branching

    C. Versioning

Further reading

- Read more about branch protection rules for GitHub: `https://docs.github.com/en/repositories/configuring-branches-and-merges-in-your-repository/managing-protected-branches/managing-a-branch-protection-rule`

- Read more about branch policies for Azure DevOps: `https://learn.microsoft.com/en-us/azure/devops/repos/git/branch-policies?view=azure-devops&tabs=browser`

13
Deepening ALM Knowledge

This last chapter will briefly explain other areas you could explore to deepen your knowledge of more advanced **Application Lifecycle Management (ALM)** approaches.

Those areas include the following:

- Testing
- Parameters for import configuration
- Third-party solutions
- Complex components
- Deployment of Azure components

Testing

One of the most important areas in which to deepen your knowledge to improve ALM is testing.

In general, when we are talking about testing, we usually mean automated tests. This means tests that are usually written in some programming language to be executed in an automated way to test the current implementation.

Every test follows a certain schema. They usually all start with some preparation, followed by the execution of the part of the implementation that should be tested, and, lastly, an assert part where we check whether the tested code does what we expect.

> **Test-driven development**
>
> There is a development approach dedicated to testing called **Test-Driven Development (TDD)**. The idea is that before any actual code is written, tests are written to see whether the code works as expected. They would all fail in the beginning since there hasn't been any code written for the implementation. Then, over time, as the software is implemented, the tests will start succeeding.

As we mentioned earlier, there are different types of tests one could create. Examples of these types are covered in the following subsections.

Unit tests

A **unit test**, as the name suggests, tests the smallest unit of an implementation. In the best case, you would write a unit test for every function. When the function is executed within the test, it should not interact with anything outside of the current test. This means everything else needs to be "mocked" away.

> Mocking
>
> When it comes to testing, "mocking" is when we recreate surrounding parts in memory. To mock, for example, Dataverse, there are different open source and paid tools available. The most well-known example is FakeXrmEasy.

This makes sure that the surrounding parts work as we expect them to work and is the only valid way of really telling whether a possible error is rooted in our implementation.

Unit tests are the minimum that should be implemented for every project.

Integration tests

The intention of **integration tests**, in contrast to unit tests, is to test whether our implementation is interacting with surrounding systems as expected.

This means surrounding systems will not be mocked away or will just be partly mocked away. For example, there could be different tests running against the same part of our implementation but mocking away different parts. This would ensure that we detect exactly which integration isn't working as expected.

Integration tests are most effective when used in combination with unit tests, as this ensures that any potential errors are not due to a problem in our part of the implementation, but rather in the communication or integration between different parts.

Performance tests

In addition, you could run **performance tests**. These are intended to run a lot of requests against software and see how the software performs.

This can be done regularly, and over time, a graph can be created of how the performance has evolved and whether it is getting better or worse.

Load tests

Load tests intend to find out at which load a software "breaks" or test whether the software is able to handle a certain expected load threshold.

They are very similar to performance tests but are mostly intended to find the breaking point. To achieve this, they usually run the tests at a higher frequency.

UI tests

UI tests run against the UI of software and test everything from end to end. For Power Platform, they could, for example, log in to a model-driven app and execute actual tasks to check whether the UI works as expected. This could be whether fields are shown/hidden depending on some values selected, the role of a certain user, or any other condition the project might have.

These tests are usually long-running and should not be executed in your normal pipeline process. Normally, they are run on a schedule, for example, once a night.

Smoke tests

Smoke tests check whether a new build of software is ready to be tested. They are run at the beginning of a test cycle.

Regression tests

Regression tests are used to test whether a major change had an impact on existing functionality. Often, they are a subset of tests that can be run against the environment after a deployment to briefly check everything in the system.

Parameters for import configuration

The import step of the Microsoft Build Tools has a lot of options for configuration. Different combinations of these options can improve the import time or change the behavior completely.

It is important to know the impacts of those configurations. Therefore, this is an area I would recommend diving more into.

As soon as you have understood the impacts, another recommendation is to add some of them as parameters to your release pipeline. With that, you can change the behavior of your pipeline whenever you start a new run. This could be, for example, whether you would like to execute an update or upgrade (as discussed in *Chapter 4*). The idea is that we can change the behavior without needing to actually change the YAML.

Third-party solutions

Another area that is often present in projects but mostly handled manually is third-party solutions.

It might be of interest to test out how they could be updated automatically.

I usually store the current version of those external solutions in my repository. In addition, I have a comma-separated list of solution names that I loop over at the beginning of my release pipeline. With the mentioned "skip lower version" switch (we learned about that switch in *Chapter 5*) of the import step, nothing will happen if the version has not been changed, but if it has changed, the new version will automatically be installed.

Complex components

As briefly mentioned in *Chapter 2*, there are certain components that require some additional steps when it comes to ALM. Those are, for example, service endpoints, where the secret isn't exported and needs to be added after every deployment, or SLAs, which get turned off with every deployment.

If some components that require extra steps are used in a solution, you should be aware of it and know how to automate those steps. Most of the time, this means a custom PowerShell script to execute some additional steps after the import is successful.

> **Power DevOps Tools**
>
> Power DevOps Tools by Wael Hamze has steps for some of those scenarios. This means you don't have to implement the PowerShell script yourself and can rather use an existing solution to the problem.

Deployment of Azure components

In nearly every project, some Azure components are included. This could be, for example, Azure Functions, Logic Apps, or a service bus.

If a project contains a component that is not directly included in Power Platform, you should know how to build and deploy it.

Bicep

When it comes to Azure components, they can even be created automatically. **Bicep** is a language with declarative syntax to deploy Azure infrastructure. In a Bicep file, you can define the infrastructure needed. The file is then used to repeatedly deploy the infrastructure in question in an automated way.

This can be used to create resources in Azure when the rest of your Power Platform implementation is deployed to a downstream environment as well. It takes away another manual step and eases deployments.

Best practices

Let's also talk about some of the best practices when it comes to ALM.

Everything should be managed

Every environment that is not dev (or hotfix) should be handled with managed solutions.

Use as few solutions as possible

It is recommended to use as few solutions as possible to minimize the management work needed.

Use only changed components

Solutions should only contain changed components.

Have at least dev, test, and prod

When it comes to environments, we should always have at least development, test, and production. In some cases, there might be a need for some additional environments too.

Use an evaluation environment for early access

We should use a separate evaluation environment to test early access features.

Create connection references manually

We should create connection references manually to be able to set the correct schema name.

As few connection references as possible

It is recommended to use as few connection references as possible. In the best case, we only have one connection reference per connector.

Use tools provided by Microsoft where possible

Where possible, it is best to use the tools provided by Microsoft. This is because they are usually supported and will be updated as the platform evolves.

Final thoughts

ALM is an area in Power Platform that is getting more and more important. A lot of companies have realized it and are now on their way to figuring out how to implement this. You have taken the first step in reading this book, which has given you good insights and knowledge about how to create a robust ALM process.

The journey does not end here, though. Like everything else on this platform, even the ALM part is changing rapidly. This means you have to be at the forefront of changes to be able to keep the ALM process working and optimized.

There are a lot of different ways of staying up to date. The ones I would most recommend are testing, building a network, and engaging with the community.

Test, test, and test

You learn most by testing and trying. You should try out new functionality or new areas whenever possible, even if you don't have a real-life project it could be applied to at the moment. If you know how something works, identifying a real-life scenario where you could apply it gets a lot easier.

Build a network

Try to attend events, especially sessions about ALM. The speakers usually keep themselves up to date and will be happy to share what they have learned.

Don't just attend the sessions but be engaged by asking questions. Catch the speaker after the session to build a network of knowledgeable people about the topic you are trying to improve on.

With that, you will always have someone to reach out to.

Community

Try to be active in the community. It does not matter whether it is a local user group or an online community. The important aspect is asking questions and trying to answer the questions of others. By doing that, you will learn a lot from the questions and answers you write and receive.

With that said, I would like to thank you for reading this book. I hope I was able to enlighten you a bit and start a passion when it comes to ALM.

Assessments

Chapter 1, An Intro to ALM

Move along – nothing to see here...haha, jk

Chapter 2, ALM in Power Platform

1. C. Configuration
2. B. Release plan
3. A. Most changes are not done in code

Chapter 3, Power Platform Environments

1. A. Development, test, production
2. C. Development
3. B. Mission critical

Chapter 4, Dataverse Solutions

1. A. Managed
2. A. Solution context menu, C. Ribbon menu
3. B. Remove from Solution

Chapter 5, Power Platform CLI

No questions to add answers to... :P

Chapter 6, Environment Variables, Connection References, and Data

1. B. 6
2. C. Data Migration Utility
3. B. They are user scoped

Chapter 7, Approaches to Managing Changes in Power Platform ALM

1.   B. Dependency to certain environments is minimized

2.   A. Main

3.   A. It is easy to manage

Chapter 8, Essential ALM Tooling for Power Platform

1.   B. Azure DevOps

2.   C. Pipelines in Power Platform

Chapter 9, Project Setup

1.   A. Development and C. PipelineDefinition

2.   A. Contribute Allow

3.   B. Project-wide

Chapter 10, Pipelines

1.   A. Power Platform Pipelines

2.   C. Download Artifact

3.   A. Pack Solution

Chapter 11, Advanced Techniques

1.   A. Variables/parameters

2.   B. Parameter

3.   C. A variable

Chapter 12, Continuous Integration/Continuous Delivery

1.   B. Branch policy

2.   C. A version of GitHub Flow

3.   A. Solution Checker

Index

‹packt›

packtpub.com

Subscribe to our online digital library for full access to over 7,000 books and videos, as well as industry leading tools to help you plan your personal development and advance your career. For more information, please visit our website.

Why subscribe?

- Spend less time learning and more time coding with practical eBooks and Videos from over 4,000 industry professionals

- Improve your learning with Skill Plans built especially for you

- Get a free eBook or video every month

- Fully searchable for easy access to vital information

- Copy and paste, print, and bookmark content

Did you know that Packt offers eBook versions of every book published, with PDF and ePub files available? You can upgrade to the eBook version at packtpub.com and as a print book customer, you are entitled to a discount on the eBook copy. Get in touch with us at customercare@packtpub.com for more details.

At www.packtpub.com, you can also read a collection of free technical articles, sign up for a range of free newsletters, and receive exclusive discounts and offers on Packt books and eBooks.

Other Books You May Enjoy

If you enjoyed this book, you may be interested in these other books by Packt:

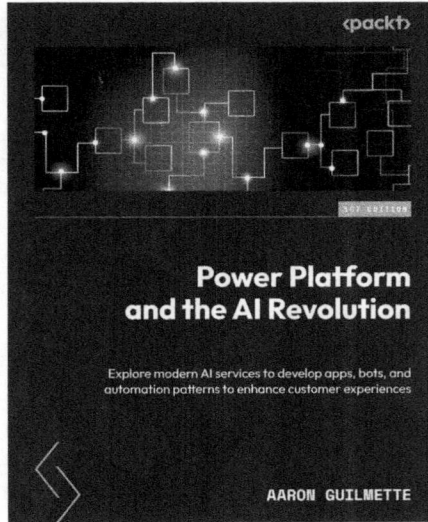

Power Platform and the AI Revolution

Aaron Guilmette

ISBN: 978-1-83508-636-0

- Interact with ChatGPT using connectors and HTTP calls
- Train AI models to identify the key elements of documents
- Use generative AI to answer questions about organizational content
- Leverage AI image recognition services to describe pictures
- Use generative AI tools to help build workflows and apps
- Build chatbots using the new Copilot Studio
- Analyze customer feedback using AI sentiment analysis tools such as AI Builder

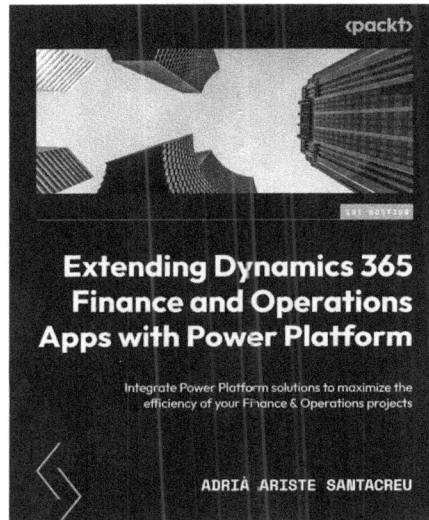

Extending Dynamics 365 Finance and Operations Apps with Power Platform

Adrià Ariste Santacreu

ISBN: 978-1-80181-159-0

- Get to grips with integrating Dynamics 365 F&O with Dataverse
- Discover the benefits of using Power Automate with Dynamics 365 F&O
- Understand Power Apps as a means to extend the functionality of Dynamics 365 F&O
- Build your skills to implement Azure Data Lake Storage for Power BI reporting
- Explore AI Builder and its integration with Power Automate Flows and Power Apps
- Gain insights into environment management, governance, and application lifecycle management (ALM) for Dataverse and the Power Platform

Packt is searching for authors like you

If you're interested in becoming an author for Packt, please visit `authors.packtpub.com` and apply today. We have worked with thousands of developers and tech professionals, just like you, to help them share their insight with the global tech community. You can make a general application, apply for a specific hot topic that we are recruiting an author for, or submit your own idea.

Share Your Thoughts

Now you've finished *Application Lifecycle Management on Microsoft Power Platform*, we'd love to hear your thoughts! Scan the QR code below to go straight to the Amazon review page for this book and share your feedback or leave a review on the site that you purchased it from.

`https://packt.link/r/1835462324`

Your review is important to us and the tech community and will help us make sure we're delivering excellent quality content.

Download a free PDF copy of this book

Thanks for purchasing this book!

Do you like to read on the go but are unable to carry your print books everywhere?

Is your eBook purchase not compatible with the device of your choice?

Don't worry, now with every Packt book you get a DRM-free PDF version of that book at no cost.

Read anywhere, any place, on any device. Search, copy, and paste code from your favorite technical books directly into your application.

The perks don't stop there, you can get exclusive access to discounts, newsletters, and great free content in your inbox daily

Follow these simple steps to get the benefits:

1. Scan the QR code or visit the link below

https://packt.link/free-ebook/9781835462324

2. Submit your proof of purchase
3. That's it! We'll send your free PDF and other benefits to your email directly

www.ingramcontent.com/pod-product-compliance
Lightning Source LLC
Chambersburg PA
CBHW081103220326
41598CB00038B/7204